Lecture Notes in Biomathematics

Managing Editor: S. Levin

73

Y. Cohen (Ed.)

Applications of Control Theory in Ecology

Proceedings of the Symposium on Optimal Control Theory
held at the State University of New York, Syracuse,
New York, August 10–16, 1986

Springer-Verlag

Berlin Heidelberg New York London Paris Tokyo

Mathematics Subject Classification (1980): 49 A 10, 49 A 22; 92-06, 92 A 15, 92 A 17

ISBN-13: 978-3-540-18104-0   e-ISBN-13: 978-3-642-46616-8
DOI: 10.1007/978-3-642-46616-8

Library of Congress Cataloging-in-Publication Data. Symposium on Optimal Control Theory (1986: State University of New York, Syracuse, N.Y.) Applications of control theory in ecology. (Lecture notes in biomathematics; 73) Bibliography: p. 1. Ecology—Mathematical models—Congresses. 2. Control theory—Congresses. I. Cohen, Yosef. II. Title. III. Series. QH541.15.M3S896 1986 574.5'0724 87-20615
ISBN-13: 978-3-540-18104-0

2146/3140-543210

ACKNOWLEDGMENTS

This volume contains the proceedings of the symposium on Optimal Control Theory, which was part of the IV International Congress of Ecology, 71st Annual Meeting of the Ecological Society of America, and 5th Meeting of the International Society of Ecological Modelling, held 10-16 August, 1986, at the State University of New York, Syracuse, New York.

Partial support for my editorial work was provided by The University of Minnesota Agricultural Experiment Station and The Department of Fisheries and Wildlife, University of Minnesota. The meticulous proof-reading by Pammela McInnis is greatly appreciated. Norma Essex helped with secretarial work and Jeff Stone helped with setting up the figures.

PREFACE

Control theory can be roughly classified as deterministic or stochastic. Each of these can further be subdivided into game theory and optimal control theory. The central problem of control theory is the so called constrained maximization (which--with slight modifications--is equivalent to minimization). One can then say, heuristically, that the major problem of control theory is to find the maximum of some performance criterion (or criteria), given a set of constraints. The starting point is, of course, a mathematical representation of the performance criterion (or criteria)--sometimes called the objective functional--along with the constraints. When the objective functional is single valued (i.e., when there is only one objective to be maximized), then one is dealing with optimal control theory. When more than one objective is involved, and the objectives are generally incompatible, then one is dealing with game theory.

The first paper deals with stochastic optimal control, using the dynamic programming approach. The next two papers deal with deterministic optimal control, and the final two deal with applications of game theory to ecological problems.

In his contribution, Dr. Marc Mangel applies the dynamic programming approach, as modified by his recent work--with Dr. Colin Clark, from the University of British Columbia (Mangel and Clark 1987)[*]--to modelling the "behavioral decisions" of insects. The objective functional is a measure of fitness. Readers interested in detailed development of the subject matter may consult Mangel (1985). My contributions deal with two applications of optimal control theory. The first is an application of classical optimal control theory to the reproductive strategy of plants. The second is the so called impulse control. It is applicable to situations where the controlled state (e.g., amount of food consumed) changes impulsively. In both cases the value functional is some measure of fitness, which is to be maximized. Dr. Robert McKelvey's contribution takes a hard look at the "tragedy of the commons" from a game theory perspective, using, specifically, the differential games approach. Finally, the contribution by Drs. Joel Brown and Thomas Vincent applies game theory ideas to the problem of evolutionary stable strategies (ESS). Their theory is more fully developed in the literature cited in their manuscript.

---

[*]See references in relevant manuscripts.

There is a growing interest among ecologists in the methods of control theory. This interest is matched by that of a small group of applied mathematicians who find ecological problems interesting and challenging. I hope that this volume will further stimulate mutual interest.

Y. Cohen
St. Paul, Minnesota
February 1987

TABLE OF CONTENTS

MODELLING BEHAVIORAL DECISIONS OF INSECTS

Marc Mangel

*Departments of Agricultural Economics,
Entomology, and Mathematics**
University of California
Davis, California 95616 USA*

Abstract.   Many problems in animal behavior can be viewed as dynamic control problems. In this paper, the dynamic viewpoint is used to analyze certain behavioral decisions of insects, in particular oviposition site selection and clutch size. The theory is motivated by work on parasitic wasps and fruit parasitic insects--particularly apple maggot (*Rhagoletis pomonella*) and medfly (*Ceratitis capitata*). The theory presented in this paper is based on Markovian decision processes in either discrete or continuous time. In these decision processes, the objective functional is a measure of fitness obtained through egg production. The paper closes with some speculations about how insects may be able to solve dynamic programming problems.

§1. Introduction: Experimental and Theoretical Motivation

The behavioral ecology of insects provides a wealth of motivation for ecological modelling. The theoretical work which is described in this paper, in particular, is motivated by a number of different sets of experiments and analyses on different types of insects. These will now be briefly described.

---

*Address correspondence to the Department of Mathematics.

Parasitic Wasps. Charnov and Skinner (1984, 1985) and Skinner (1985) studied clutch size as a function of host volume in the wasp *Nasonia*. In their analysis, they found that most observed field clutch sizes were smaller than the clutch size which maximized fitness per host; this will be called the "Lack Clutch Size" (or LCS) in analogy to David Lack's work on clutch size in birds. Charnov and Skinner provide an explanation for the apparent "non-optimality" of the observed clutches in terms of rates, in analogy to the marginal value theorem of Charnov (1976). It will be seen here that what one means by "optimal," in fact, drives what is considered non-optimal behavior.

Apple Maggot. R. Prokopy and his collaborators and students, particularly B. Roitberg have conducted an elegant series of experiments on the field behavior of *Rhagoletis pomonella* (Prokopy and Roitberg 1984, Roitberg 1985, Roitberg and Prokopy 1981,1982,1983,1984, Roitberg et al. 1982,1984).

Roitberg and Prokopy (1983) studied the effect of host deprivation on the response of *R. pomonella* to its oviposition marking pheromone (OMP). They observed the following kinds of results:

Time Since Last Oviposition (min.)	Percent of Flies Ovipositing in OMP-Marked Fruit
5	10%
10	45%
20	65%
40	66%
80	85%

These numbers were read off by eye from Figure 1 of Roitberg and Prokopy (1983), so they might be a tad off; the trend is clear however. They conclude: "Thus, acceptance of parasitized hosts by short term (< 96 h), host-deprived flies must be due to changes in physiological state associated with host deprivation" (Roitberg and Prokopy 1983, page 71). The exact nature of these physiological changes remain to be determined, but I will provide some speculations of my own in §5.

In another paper on the foraging behavior of the apple maggot (Roitberg et al. 1982), a study is made of the relationship between the number of fruit clusters on a tree, the residence time of the fly in the tree and the giving up time (GUT), which is defined here as the time since the last oviposition before emigration from the tree. The observed data are (read off from Figures 12 and 13 of Roitberg et al. 1982):

Number of Fruit Clusters	Residence Time (min.)	GUT (min.)
2	20	12
4	55	13
8	70	7
16	75	4

(Note: these are only approximate values since they are read off the figures by eye and correspond to mean values.) Note that the GUT decreases with residence time in the patch; this is an observation which is difficult to explain in terms of classical foraging theory.

Medfly. My colleague, J. Carey and his students recently developed an artificial host for field and laboratory studies of the behavior of another tephritid fly, the Mediterranean fruit fly *Ceratitis capitata*. His student, R. Freeman, studied the distribution of eggs as a function of host volume (it is difficult to determine "clutch size" in this case, but total number of eggs is easy). They find the kind of results shown in Figure 1. The key observation here is that the number of eggs per host levels off with host volume. Although it is possible, it appears that this effect is more than the medflies simply running out of eggs.

The theoretical motivation for this paper is recent work done in conjunction with Colin Clark on the theory of foraging (Mangel and Clark 1986). This theory is based on the use of Markovian decision processes for modelling of foraging actions and decisions. In the next section, this approach is described in more detail. It is applied, in the third section, to a model for the behavior of parasitic wasps (also see Mangel 1987) and, in the fourth section to a model for the rose hips fly *Rhagoletis basiola*. The second model is currently under further development and will be used to analyze field experiments by B. Roitberg. The fifth section contains a discussion and, in the spirit of a workshop, some speculations. In particular: how do organisms solve stochastic dynamic programming problems?

§2. Markovian Decision Models

In this section, an approach to modelling behavioral decisions based on Markovian decision processes is described. Mangel and Clark (1986) call this approach "unified foraging theory" since it allows one to treat the three main aspects of behavior--finding food, avoiding predation, and reproducing--in a unified manner.

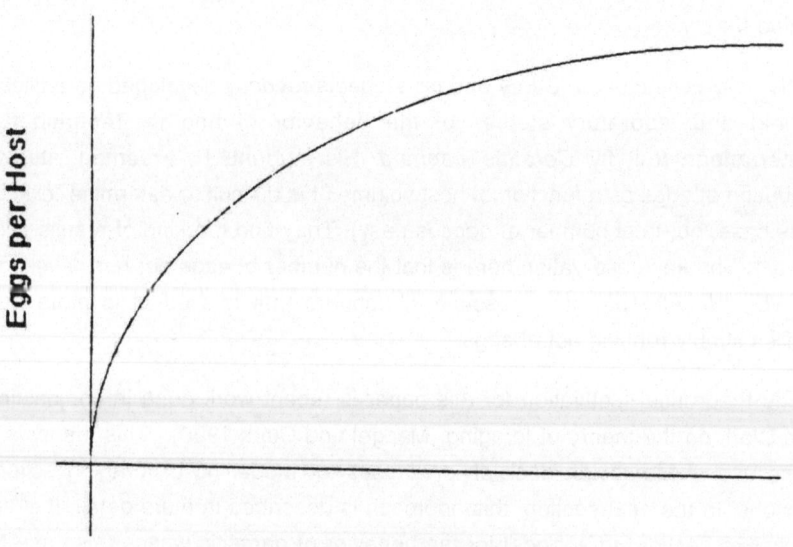

Figure 1. Results of Carey and Freeman on Medfly response to host volume.

The basic objective of these decision models is to be able to treat trade-offs in a consistent manner. One is thus able to deal with a wide variety of ethological problems (see, e.g., Huntingford 1984). There are three main components to this theory:

1) A state variable (variables), $X(t)$, which provides a means of assessing the current state of the organism. For example, for insects which produce a fixed number of eggs, $X(t)$ could represent the number of eggs remaining at time t. The state variable changes over time, subject to physiological constraints (it will be seen that these are very important for insect decisions), decisions by the organism, and (usually random) environmental effects. The state variable provides a means for connecting the physiological state of the organism with its behavior.

2) An objective functional, which depends upon the current value of the state variable and provides a measure of "value" for the current state variable when assessed at a later point in time. This objective provides a way for connecting long term and short term behaviors. For example, for insects the objective functional might be lifetime fitness obtained through egg production, given that the initial number of eggs is $X(0) = x$.

3) A methodology for optimizing the objective functional, subject to the stochastic dynamics of $X(t)$ and any appropriate constraints. The method from stochastic control used in this paper is stochastic dynamic programming (Aoki 1967, Bertsekas 1976, Mangel 1985). This method is actually little more than a bookkeeping technique (with probability 1 something will happen!) and clever use of computers.

The theory presented here, and in Clark (1987), Mangel (1987) and Mangel and Clark (1986), is easy to use. It involves parameters which should be easily measured in the field or laboratory; the mathematical formulation is straightforward and the required computations can be implemented on a desk top microcomputer.

§3. Clutch Size in Parasitic Insects

The results presented in this section were motivated by the work of Charnov and Skinner (1984,1985) and Skinner (1985). There is a considerable overlap between this section (summarized from Mangel 1987) and their papers, as well as with those of Iwasa et al. (1984) and Parker and Courtney (1983). The objective of this section is to show, by means of the simplest possible model, how the state variable approach can be applied and used to understand clutch size decisions.

Imagine an insect which starts its life with a reserve $X(0) = X_0$ of mature eggs and attacks one of H different kinds of hosts. If it encounters a host of type i and lays a clutch of size ϕ in this host, then assume that its lifetime fitness increases by $W_i(\phi)$. The computation of the increment in fitness, $W_i(\phi)$, is a very nontrivial matter and Charnov and Skinner (1984) and Skinner (1985) did a great job of developing Ws for a number of different insect-host systems.

The state variable, $X(t)$, is defined by

[3.1] $X(t)$ = number of eggs remaining at time t.

The dynamics of the state variable are then quite straightforward. Consider a small interval of time Δt and define

[3.2] $\lambda_i \Delta t$ = Probability of encountering a host of type i in (t , t+Δt) .

If a clutch of size ϕ_i (yet to be determined) is laid on a host of type i, then [3.1] and [3.2] give

[3.3] $X(t + \Delta t) = X(t) - \phi_i$ with probability $\lambda_i \Delta t$.

The objective functional comes next. In order to define it, let T denote the maximum possible lifetime of the insect and set

[3.4] $F(x , t , T)$ = maximum expected fitness from egg
                             production between t and T when
                             the number of eggs remaining at t
                             is $X(t) = x$.

Then one certainly has

[3.5] $F(x, T, T) = 0$

since there is no value to having any eggs at death. In order to develop the dynamic programming equation (DPE) that $F(x, t, T)$ satisfies, first define $\rho(t)$ by

[3.6] $1 - \rho(t)\Delta t$ = Probability of surviving to time $t+\Delta t$, given
                    that the insect is alive at time t ;

this quantity can be computed from the usual survivorship curves of demography (see, e.g. Mangel 1987).

Now reason as follows: If the insect encounters a host of type i it can lay a clutch of any size between 0 and the current number of remaining eggs. If the clutch size is ϕ, then the increment in fitness is $W_i(\phi)$ and the number of remaining eggs is decreased by ϕ. In symbols, one has

[3.7] $$F(x, t, T) = \sum_{i=1}^{H} \lambda_i \, \Delta t \, \max_{\phi \leq x} \left\{ W_i(\phi) + (1 - \rho\Delta t) \, F(x-\phi, t+\Delta t, T) \right\}$$

$$+ \left(1 - \sum_{i=1}^{H} \lambda_i \, \Delta t\right) (1 - \rho\Delta t) \, F(x, t+\Delta t, T) .$$

Equation [3.7] is the basic DPE associated with this problem. Note how the constraint on the number of eggs arises in a most natural way. There are two ways to analyze equation [3.7]. The first is to set $\Delta t = 1$. In that case, time is measured in discrete units (although the specific unit of time is not given, so that it could be quite small). Equation [3.7] becomes

[3.8] $\quad F(x,t,T) = \displaystyle\sum_{i=1}^{H} \lambda_i \; \max_{\phi \le x} \left\{ W_i(\phi) + (1-\rho)\, F(x-\phi,t+1,T) \right\}$

$$+ \left(1 - \sum_{i=1}^{H} \lambda_i\right)(1-\rho)\, F(x,t+1,T) \; .$$

This equation is easily iterated backwards, starting at $t = T-1$, on a desk-top microcomputer. Its solution leads to a number of predictions which will be discussed shortly.

The alternative to $\Delta t = 1$ is the continuous time limit in which Δt approaches 0. To deal with this case, Taylor expand [3.7] in powers of Δt to obtain

[3.9] $\quad F(x,t,T) = \displaystyle\sum_{i=1}^{H} \lambda_i \Delta t \; \max_{\phi \le x} \left\{ W_i(\phi) + F(x-\phi,t,T) + O(\Delta t) \right\}$

$$+ F(x,t,T) + \frac{\partial F}{\partial t}\Delta t - \left(\rho + \sum_{i=1}^{H} \lambda_i\right)\Delta t\, F(x,t,T) + o(\Delta t)$$

where $O(\Delta t)$ and $o(\Delta t)$ represent quantities such that

[3.10]
$$\lim_{\Delta t \to 0} \frac{O(\Delta t)}{\Delta t} = \text{constant}$$
$$\lim_{\Delta t \to 0} \frac{o(\Delta t)}{\Delta t} = 0 \; .$$

Dividing by Δt and letting Δt approach 0 gives the equation

[3.11] $\quad -\dfrac{\partial F}{\partial t} = \displaystyle\sum_{i=1}^{H} \lambda_i \max_{\phi \le x} \left\{ W_i(\phi) + F(x-\phi,t,T) \right\} - \left(\rho + \sum_{i=1}^{H} \lambda_i\right) F(x,t,T) \; .$

This is a nonlinear, partial differential-difference equation. It is much harder to solve than the discrete time version [3.8]. Some techniques for solving such equations are discussed by Ahmed and Teo (1981) and Teo and Wu (1984).

Returning to the discrete time version [3.8], one finds the following predictions arising from the solution (see Mangel 1987 for more details).

P_1: Older insects should be less choosy about where they lay their eggs. For example, there should be more superparasitism near death.

P_2: A cohort of identical insects which start life together will, at later times, have a distribution on the values of the number of eggs remaining (caused by weather, food, host encounters, etc.). This will lead to a distribution in clutch sizes.

P_3: As the chance of finding hosts for which the optimal clutch size ϕ^* is larger increases, the observed distribution of clutches should change so that smaller clutches are more frequent.

P_4: As the conditional probability of survival decreases, the frequency of large clutches should increase.

P_5: As the time horizon $T - t$ decreases, for example by host deprivation, the frequency of larger clutches should increase.

Some of these predictions (e.g. P_1, P_2, P_4) can be seen by qualitative examination of the dynamic programming equation [3.8]. Others (e.g. P_3, P_5) are less obvious--it helps to solve [3.8] to see them--but are easily understood when one starts thinking in the paradigm that UFT provides. It is the paradigm of constrained, Markovian dynamics which guides the prediction.

§4. A model for *Rhagoletis basiola*

A model for *Rhagoletis basiola* is described in this section. It differs in many ways from the model of the previous section. The differences are based on a number of biological details. The most important are these: upon encountering a host fruit, the fly either lays one egg or no eggs. The fitness accrued to the mother from this egg depends upon whether or not the host was previously parasitized and if so, when. For example, one could develop the following kind of data (B.   Roitberg, personal communication):

i	Host Type	Relative Fitness, W_i
1	Unparasitized	2.00
	Previously parasitized	
2	1 day before	1.50
3	2 days before	1.00
4	3 days before	0.75
5	4 days before	0.50
6	Larva present (host parasitized 5 or more days previously)	0.20

The lifetime of the fly is about 15 days; each day is divided into 20 hours in which the fly does activities other than search for oviposition sites and 4 hours in which it searches for oviposition sites. A "timeline" for each day can be developed as follows:

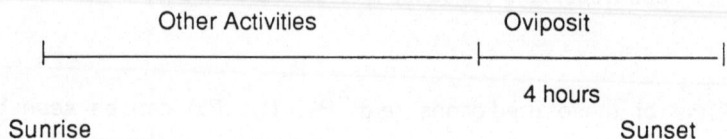

Finally, in this model there are two state variables defined as follows:

X(t , D) = number of mature eggs being held at the start of period t on day D

[4.1]

Y(t , D) = number of oocytes (potential eggs), remaining at the start of period t on day D.

The variable $X(t, D)$ has a capacity constraint such that

[4.2] $X(t, D) \leq C$

where C is the maximum number of mature eggs that the fly can hold at any time. The variable $Y(t, D)$ satisfies the constraint that

[4.3] $Y(t, D) \leq Y_M$

where Y_M is the maximum number of potential eggs.

Now define a fitness function $F_D(x, y, t, T)$ as follows:

[4.4] $F_D(x, y, t, T) =$ maximum expected fitness through egg production when D days remain, when $X(t, D) = x$, $Y(t, D) = y$, and $T - t$ is the number of periods remaining in day D.

That is, T is the time horizon for foraging for oviposition sites within a day. When analyzing [4.4], it will be understood that $D = 1$ corresponds to the last day of the fly's life. Thus one has the end condition

[4.5] $F_1(x, y, T, T) = 0$.

The end condition connecting day D and $D - 1$ is more complicated. Suppose that a fly ends day D with x mature eggs and y potential eggs remaining. During the night, it can, in principle, mature enough eggs to reach the capacity C. Thus, it starts the next day with C mature eggs, as long as $y \geq C - x$. Otherwise it starts the next day with $x + y$ mature eggs. Thus, in addition to [4.5], one has

[4.6]
$$F_D(x,y,T,T) = \begin{cases} F_{D-1}(C, y-C+x, 0, T) & \text{if } y \geq C-x \\ F_{D-1}(x+y, 0, 0, T) & \text{if } y < C-x. \end{cases}$$

Now consider the dynamic programming equation for behavior within a given day. The discrete time formulation will be used. Let

[4.7] $1 - \rho_D(t)$ = Probability that the fly is alive at the start of period t+1 given that it is alive at the start of period t with D days remaining.

Also assume that the length of a period is sufficiently great that an egg can be matured in a period, if any potential eggs remain. Finally, introduce the following notation

[4.8] λ_i = Probability of encountering a host of type i in a period (note: λ_i could easily be a function of D and t with no change in the algorithm)

and the "indicator functions"

$$I_{x<C} = \begin{cases} 1 & \text{if } x < C \\ 0 & \text{if } x = C \end{cases}$$

$$I_{y>\dot{y}} = \begin{cases} 1 & \text{if } y > \dot{y} \\ 0 & \text{if } y \leq \dot{y} \end{cases}$$

Finally, set

$$\lambda_0 = 1 - \sum_{i=1}^{H} \lambda_i$$

and $W_0 = 0$. With this notation, the same kind of logic that leads to [3.8] leads to the following dynamic programming equation

$$[4.9] \quad F_D(x, y, t, T) = \sum_{i=1}^{H} \lambda_i \max \Big[(1 - p_D(t)) \{ F_D(x+1, y-1, t+1, T) I_{x<C} I_{Y>0}$$

$$+ F_D(x, 0, t+1, T) I_{x<C} (1 - I_{Y>0}) + F_D(C, y, t+1, T) (1 - I_{x<C}) \} ;$$

$$W_i + (1 - p_D(t)) \{ F_D(x, y-1, t+1, T) I_{Y>0} + F_D(x-1, 0, t+1, T) (1 - I_{Y>0}) \} \Big].$$

Although it looks formidable, equation [4.9] is no harder to solve than [3.8]--it's just that the indicator functions make it look more complicated.

Equation [4.9] is somewhat complex, but it is easily solved on a desk-top microcomputer. More interestingly, one can develop Monte Carlo simulations in which insects behave "optimally" according to the solution of [4.9] but encounter host types randomly. By using the simulation, one can perform "computer experiments" analogous to the field and lab experiments on real flies. (Mangel 1987 provides a further discussion of such computer experiments.) For example, a simulation was programmed for the following situation: 100 flies start the last day of their lives, which lasts for 40 periods, with 2 mature eggs and 14 potential eggs. They encounter the six host types randomly, each with equal probability, and make oviposition decisions according to equation [4.9]. Using this simulation, one can perform "host deprivation" experiments by reducing the time horizon. Figure 2 shows the results of such an experiment in which one sees an increase in the oviposition rate in either marked fruit (upper panel) or very inferior hosts (lower panel) with increased host deprivation (which for this problem corresponds to decreased time horizon). This pattern compares very well with the results of Roitberg and Prokopy (1983) discussed previously.

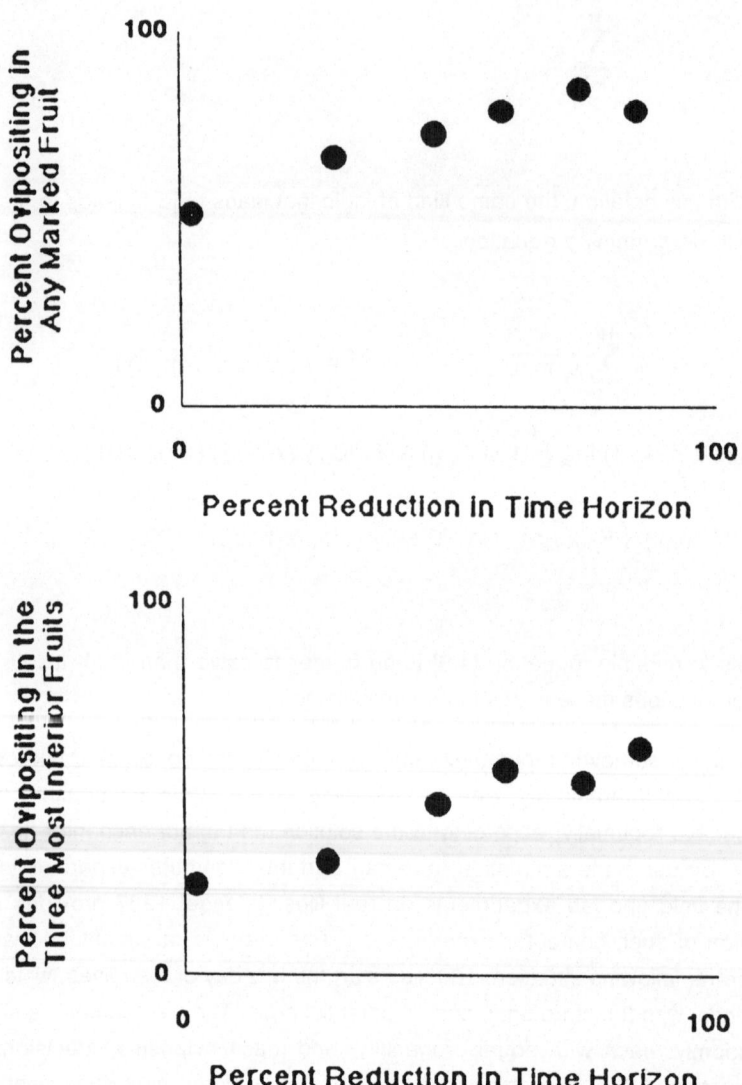

Figure 2. Results of simulation experiments on host deprivation. (a) Fraction of flies ovipositing in any marked fruit. (b) Fraction of flies ovipositing in any of the three most interior fruits.

§5. Discussion and Speculation

The two models presented in this paper provide examples of how Markov decision theory based on state variable models can be used to analyze insect behavior. Other models are possible as well. For example, one can take into account the energetic reserves of the insect and include a state variable that measures those reserves. In such a case, the insect must choose between foraging for food and foraging for oviposition sites (see Mangel 1987 for an example). Markov decision processes thus provide a method for analyzing a wide variety of behavioral activities.

I will close this paper with speculations (fitting for conference proceedings) about how insects might solve dynamic programming problems. There are at least two lines of thought about this question. The first is based on a hypothesis of R. Fox, School of Physics at Georgia Institute of Technology:

> **Fox's Hypothesis**: If one can simulate fast enough then any stochastic optimization problem can be solved.

With this in mind, one can leap to the speculation that perhaps one reason for the evolution of molecular and chemical chaos in organisms is to provide a mechanism for simulation. Most ecologists are familiar with chaos through nonlinear population maps such as the logistic:

[5.1] $$N(t+1) = N(t) + rN(t)(1 - N(t))$$

which goes through a series of bifurcations to chaos as r increases. There are, however, many chemical systems which involve continuous time reaction kinetics that also lead to dynamical chaos. Perhaps these kinetics provide the underlying "tools" by which organisms can solve dynamic programming problems.

In addition to a "chaotic simulation" approach, there is extremely exciting work currently being done by J. Hopfield and his collaborators (e.g. Hopfield 1982, Hopfield and Tank 1985,1986) on the use of model neuron systems to solve optimization problems. They find that large, interconnected networks of model neurons can find good (if not fully optimal) solutions to hard optimization problems such as the

Travelling Salesman Problem. It is likely that their work can be adapted to deal with dynamic optimization problems such as the ones described in this paper (the Travelling Salesman Problem can, in fact, be put into a recursive dynamic form as well). Much exciting work needs to be done!

Acknowledgements. This work was partially supported by the California Department of Food and Agriculture through contract 6822, by NSF Grant BSR 86-1073, and by the Agricultural Experiment Station of the University of California. I thank Jim Carey, Colin Clark, Hugh Dingle, Bob Dowell, David Foote, Rachel Freeman, Mike Hassell and Bernie Roitberg for helpful discussions. Bernie Roitberg and Yossi Cohen both carefully read the manuscript and provided numerous improvements.

Literature Cited

Ahmed, N.U., and K.L. Teo. 1981. *Optimal Control of Distributed Parameter Systems* . North Holland, New York.

Aoki, M. 1967. *Optimization of Stochastic Systems.* Academic Press, New York.

Bertsekas, D. 1976. *Dynamic Programming and Stochastic Control.* Academic Press, New York.

Charnov, E.L. 1976. Optimal foraging: the marginal value theorem. *Theoretical Population Biology* 9: 129-136.

Charnov, E.L., and S.W. Skinner. 1984. Evolution of host selection and clutch size in parasitoid wasps. *Florida Entomologist* 67: 5-21.

Charnov, E.L., and S.W. Skinner. 1985. Complementary approaches to the understanding of parasitoid oviposition decisions. *Environmental Entomology* 14: 383-391.

Clark, C.W. 1987. The lazy, adaptive lion. *Animal Behavior.* In press.

Hopfield, J.J. 1982. Neural networks and physical systems with emergent collective computational abilities. *Proceedings of the National Academy of Sciences, USA* 79: 2554-2558.

Hopfield, J.J., and D.W. Tank. 1985. "Neural" computation of decisions in optimization problems. *Biological Cybernetics* 52: 141-152.

Hopfield, J.J., and D.W. Tank. 1986. Collective computation with continuous variables. Pages 155-170 *in* E. Bienenstock et al., editors. *Disordered Systems and Biological Organization.* Springer Verlag, Heidelberg.

Huntingford, F.A. 1984. *The Study of Animal Behavior.* Chapman and Hall, New York.

Iwasa, Y., Y. Suzuki, and H. Matsuda. 1984. The theory of oviposition strategy of parasitoids. I. Effect of mortality and limited egg number. *Theoretical Population Biology* 25: 205-227.

Mangel, M. 1985. *Decision and Control in Uncertain Resource Systems.* Academic Press, New York.

Mangel, M. 1987. Oviposition site selection and clutch size in parasitic insects. *Journal of Mathematical Biology.* In press.

Mangel, M., and C.W. Clark. 1986. Towards a unified foraging theory. *Ecology* 67: 1127-1138.

Parker, G.A., and S.P. Courtney. 1984. Models of clutch size in insect oviposition. *Theoretical Population Biology* 26: 27-48.

Prokopy, R.J. and B.D. Roitberg. 1984. Foraging behavior of true fruit flies. *American Scientist.* January-February: 41-49.

Roitberg, B.D. 1985. Search dynamics in fruit parasitic insects. *Journal of Insect Physiology* 11: 867-872.

Roitberg, B.D., and R.J. Prokopy. 1981. Experience required for pheromone recognition by the apple maggot fly. *Nature* 292: 540-541.

Roitberg, B.D., and R.J. Prokopy. 1982. Influence of intertree distance on foraging behavior of *Rhagoletis pomonella* in the field. *Ecological Entomology* 7: 437-442.

Roitberg, B.D., and R.J. Prokopy. 1983. Host deprivation influence on response of *Rhagoletis pomonella* to its oviposition deterring pheromone. *Physiological Entomology* 8: 69-72.

Roitberg, B.D., and R.J. Prokopy. 1984. Host visitation sequence as a determinant of search persistence in fruit parasitic tephritid flies. *Oecologia* 62: 7-12.

Roitberg, B.D., Cairl, R.S. and R.J. Prokopy. 1984. Oviposition deterring pheromone influences dispersal distances in tephritid fruit flies. *Entomologia experimentalis applicata* 35: 217-220.

Roitberg, B.D., van Lenteren, J.C., van Alphen, J.J.M., Galis, F., and R.J. Prokopy. 1982. Foraging behavior of *Rhagoletis pomonella*, a parasite of Hawthorn (*Crataegus_virdis*), in nature. *Journal of Animal Ecology* 51: 307-325.

Skinner, S.W. 1985. Clutch size as an optimal foraging problem for insects. *Behavioral Ecology and Sociobiology* 17: 231-238.

Teo, K.L. and Z.S. Wu. 1984. *Computational Methods for Optimizing Distributed Systems*. Academic Press.

OPTIMAL REPRODUCTIVE STRATEGIES IN ANNUAL PLANTS

Yosef Cohen

Department of Fisheries and Wildlife
University of Minnesota, St. Paul, MN 55108 USA

Abstract.   Optimal control theory is invoked to investigate the conditions under which an annual plant exhibits two distinct reproductive strategies which result in maximum fitness by the end of a fixed-length growing season. In particular, a plant which is capable of reproducing from tubers and seeds is considered. When the plant's photosynthesizing tissue and the reproductive tissues are subject to predation of differing intensity, then the optimal reproductive strategy results in channeling all of the plant's photosynthetic product to the production of one tissue type (i.e., photosynthesizing tissue, or one of the reproductive tissue types) at a time. Whether the plant will exhibit one or two reproductive strategies (seeds or tubers, or both) depends on the predation rate on the three tissue types. Factors which contribute to suboptimal growth allocation are discussed.

§1. Introduction

Evolutionary theory asserts that life history evolves so as to maximize individual fitness. One particularly interesting example is the case where annual plants, which have two distinct reproductive strategies, display either or both in some seasons and in some habitats. I analyze the case where herbivory on two types of reproductive tissue, as well as on the photosynthesizing tissue, plays a role in the optimal allocation of photosynthetic products to the growth of the appropriate reproductive tissue during a fixed-length growing season. The central question is how predation rates on those three tissue types (two reproductive tissues and one photosynthesizing tissue) and the fitness resulting from achieving a certain biomass of these tissues by the end of the growing season affect the optimal growth trajectory and reproductive strategy of an individual plant. To address this question, the powerful methods of optimal control theory (Leitman 1981) are invoked. A similar method was applied to the analysis of optimal plant growth and reproduction by Vincent (1979), Vincent and Pulliam (1980), King and Roughgarden (1982a,b), and Iwasa and Roughgarden (1984).

Consider the submerged aquatic macrophytes, *Potomageton pectinatus*, as an example. It is common in many parts of the world and is particularly abundant in wetlands in Minnesota, North Dakota, Ontario and Manitoba (Anderson 1978, Aiken 1979). *P. pectinatus* displays two distinct reproductive strategies. In many parts of the prairie regions of North America it grows fast during the summer, and by the end of the short growing season (usually during the beginning of October) it does or does not produce seeds, which overwinter. After the spring thaw, it grows from tubers and seeds. Negative correlation between the mass of tubers and the mass of seeds has been reported (Yeo 1965) and observed in laboratory experiments (personal observation). When seeds are produced, they develop from flowers which grow during the middle of the growing season. Towards the end of the growing season tuber biomass increases more rapidly compared to the beginning of the growing season (Engel 1985). The tubers and seeds of *P. pectinatus* are an important food source for breeding waterfowl populations (Cottam 1939) which can consume up to 40% of its standing biomass (Anderson and Low 1976). Some waterfowl species, primarily the diving ducks, specialize on feeding on its tubers. This plant species, particularly its above ground biomass, is also heavily preyed upon by large herbivores such as moose and deer in their search for high sodium content foods (Botkin et al. 1973).

§2. The Model

Let

x_1 denote the tuber biomass,

x_2 denote the seed biomass, and

x_3 denote the photosynthesizing tissue biomass.

Assume that the remaining tissue mass (such as supportive tissue) is negligible compared to $x_p = \Sigma x_i$, $i = 1, 2, 3$, where x_p denotes the total plant biomass. The plant growth rate is a function of its photosynthesizing tissue biomass, and therefore

$$dx_p / dt = g(x_3) .$$

Denote by $u_i(t)$ the proportion of total growth due to photosynthesis that is allocated to growth of tissue of type i at time t, with $\Sigma u_i = 1$. Suppose that **within a single growing season** each tissue type is subjected to a predation rate which depends on the mass of the particular tissue, but not on the density of predators. Denote the predation rate on tissue i by $f_i(x_i)$. Now each tissue type grows according to

[2.1] $dx_i / dt = u_i(t) g(x_3) - f_i(x_i), \quad u_i \in [0, 1], \quad i = 1, 2, 3$

with

[2.2] $x_1(0) = x_{10}, \quad x_2(0) = 0, \quad$ and $\quad x_3(0) = x_{30} .$

Next, suppose that total plant fitness is a function of the biomass of the three tissue types by the end of the growing season of a **fixed** duration $[0, T]$ such that

[2.3]   Fitness $= F_1(x_1(T)) + F_2(x_2(T)) + F_3(x_3(T)) .$

I assume that the plant has evolved so that its fitness is maximized, and investigate the optimal allocation of $u_i(t)$ which maximizes [2.3] given [2.1] and [2.2]. Equations [2.1] and [2.2]--with [2.3] to be maximized--define an optimal control problem.

Next, I make and discuss general assumptions about the functions g, f, and F:

(a) $g(0) = 0$, $g'(x_3) > 0$ for $x < x^\#$, $g'(x_3) < 0$ for $x > x^\#$, $g'(x_3) = 0$ for $x = x^\#$, $g''(x_3) < 0$.

Primes denote derivatives with respect to the dependent variables. This assumption means that the plant grows fast when small, and then its growth rate slows down as it reaches a maximum growth rate at $x = x^\#$. As the plant continues to grow, its growth rate starts to decline at an increasing rate. This type of function provides for self shading and senescence, both commonly occurring in many annual plants.

(b) $f_i(x_i) > 0$, $f_i(0) = 0$, $f_i'(x_i) > 0$, for $i = 1, 2, 3$.

If $f_i'' = 0$, this assumption means that predation rate on a particular tissue increases as that tissue biomass increases with a proportion coefficient denoted by a_i. If $f_i'' < 0$, the assumption means that predation rate increases at a slowing rate as tissue i biomass increases, and provides for satiating effect.

(c) $F_i(0) = 0$, $F_i(x_i) > 0$, $F_i'(x_i) > 0$ for $x_i < x_i^\S$, $F_i'(x_i) < 0$ for $x_i > x_i^\S$, $F_i(x_i^\S) = 0$, $F_i''(x_i) < 0$ for $i = 1, 2, 3$.

This assumption means that there is a particular mix of tissue biomass which maximizes fitness.

To solve the problem of how to allocate $u_i(t)$ so as to maximize [2.3] the so called Hamiltonian is formed:

[2.4] $$H = \sum_{i=1}^{3} \lambda_i(t) [u_i(t) g(x_3(t)) - f_i(x_i(t))]$$

where the adjoint variables $\lambda_i(t)$ are defined by

[2.5] $d\lambda_i / dt = -\partial H / \partial x_i = \lambda_i(t) f_i'(x_i)$, $\lambda_i(T) = F_i'(x_i(T))$, $i = 1, 2$,

and

[2.6] $d\lambda_3 / dt = -\partial H / \partial x_3 = -g'(x_3) \sum_{i=1}^{3} \lambda_i(t) u_i(t) + \lambda_3(t) f_3'(x_3)$, $\lambda_3(T) = F_3'(x_3(T))$

The terminal conditions in [2.5] and [2.6] are derived from the so called transversality conditions (Leitman 1981).The maximum principle (Pontryagin et al. 1962) now dictates that if $u_i(t)$ is such that H is maximized throughout $[0, T]$, then the control problem posed in [2.1]-[2.3] is solved.

§3. The Adjoint Variables and the Hamiltonian

Before proceeding with the solution, let us interpret the biological meaning of the adjoint variables λ_i and the Hamiltonian H in the context of our problem. This interpretation parallels that of economic optimal control models (for details see Isard and Liossatos 1979). Iwasa and Roughgarden (1984) discuss the adjoint variables and their biological interpretation in detail. Their presentation is particularly lucid. To simplify the discussion, I shall look at the maximum principle (Pontryagin et al. 1962) with some simplifying assumptions regarding the functions discussed in this section.

Let $x(t) = [x_1(t), x_2(t), x_3(t)]^T$ and let $f = [u_1g-f_1, u_2g-f_2, u_3g-f_3]^T$. Our model then becomes

[3.1] $dx / dt = f(x, u)$, $x(0) = x_0$

where $u = [u_1, u_2, u_3]^T$. Denote by U the constraint set of u. Let

[3.2] $V(x(t)) = \sum_{i=1}^{3} F_i(x_i(T))$

where *V* is a function whose value is the maximum of the objective *Fitness* (equation [2.3]) of the control problem, given that one starts at *x(t)* at time *t*. According to the principle of optimality (Bellman 1957), whatever the initial decision is, if the remaining decisions are to be optimal, then they must result in an optimal path with regard to the initial decision. To interpret *H* and λ, consider the last short time interval in our problem (Figure 1). Our value function *V* changed from *V(x(T-δt))* to *V(x(T))*. This change is due to the value function *V(x(T))* at time *T*. The control action at *T-δt* should be chosen such that

[3.3]
$$V(x(T-\delta t)) = \max_{\substack{u(\tau) \in U \\ \tau \in [T-\delta t, T]}} \left\{ V(x(T)) \right\} .$$

In other words, one should choose *u* such that our control action during *[T-δt , T]* would maximize our current value *V(x(T-δt))*. Assuming that *V* is twice differentiable and Taylor series expanding one gets

[3.4]
$$V(x(T)) = V(x(T-\delta t)) + V_x(x(T-\delta t)) \, f \, \delta t + o(\delta t)$$

where subscripts denote partial derivatives with respect to the subscript and where *o(δt)* is a function such that

$$\lim_{\delta t \to 0} \frac{o(\delta t)}{\delta t} = 0 .$$

Therefore, from [3.3] and [3.4] one gets

[3.5]
$$V(x(T-\delta T)) = \max_{\substack{u(\tau) \in U \\ \tau \in [T-\delta t, T]}} \left\{ V(x(T-\delta T)) + V_x(x(T-\delta T)) \, f \, \delta t \right\} + o(\delta t)$$

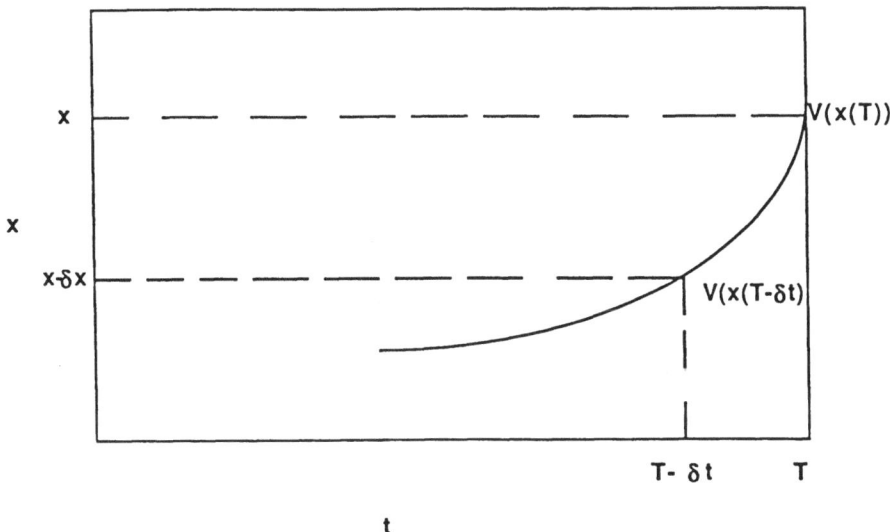

Figure 1. A sketch of change in the value function *V* with time and with *x*.

Upon taking $\delta t \rightarrow 0$ and rearranging [3.11] one obtains

$$[3.6] \qquad 0 = \max_{u(\tau) \in U} \left\{ V_x(x(T-\delta t)) \, f \right\} \qquad .$$

This is a particular case of the well known Hamilton-Jacobi-Bellman equation. Now if one defines $\lambda(t) = V_x$, then the adjoint variables λ can be interpreted as the per unit change in the objective function for a small change in x. The function $V_x \, f$ to be maximized in [3.6] is actually the Hamiltonian. $\lambda(t)$ is then the marginal fitness per unit biomass at time t. It may be called the "shadow fitness" of a unit of biomass. In particular, $\lambda(T) = F'(x(T))$ is the marginal fitness of x at time T.

Since by definition $H = \lambda dx / dt$, then $Hdt = \lambda dx$, where dx represents the change in x biomass from t to $t+dt$, when one is at x and u is applied. λdx then represents the value, in units of fitness, of the incremental biomass dx. Hdt is the contribution to our objective *Fitness* in [2.3] for t to $t+dt$ when $x(t) = x$ and $u(t) = u$. Therefore, one must seek to maximize H at each instant of time. Furthermore, since $d\lambda / dt = -H_x$, then $-d\lambda = Hxdt$ or $-d\lambda = \lambda fxdt$. In other words, along the optimal path, the decrease in the fitness value of x from t to $t+dt$, should equal the marginal increase(in fitness) from increasing x.

§4. The Optimal Solution

I now turn to the optimal solution of the problem. To determine the functions $u_i(t)$ which maximize H, isolate the terms in H which depend on u_i, define the switching function

$$\sigma(t) = g(x_3) \sum_{i=1}^{3} u_i(t) \, \lambda_i(t)$$

and look for $u_i(t)$ which maximize $\sigma(t)$ during $[0, T]$. These $u_i(t)$ will also maximize H. Thus, for $g > 0$, σ will be maximized according to the rule

$$u_i(t) = 1 \quad \text{if} \quad \lambda_i(t) = \max[\lambda_1(t), \lambda_2(t), \lambda_3(t)],$$

$u_i(t) = 0$ if $\lambda_i(t) < \max [\lambda_1(t), \lambda_2(t), \lambda_3(t)]$.

From the terminal conditions in [2.5] and [2.6] one has, at the end of the growing season, T, that:

[4.1] $\sigma(T) = g(x_3(T)) \sum_{i=1}^{3} F_i'(x_i(T)) \, u_i(T)$

[4.2] $d\lambda_i(T) / dt = F_i'(x_i(T)) \, f_i'(x_i(T))$, $i = 1, 2$

[4.3] $d\lambda_3 / dt = -g'(x_3(T)) \sum_{i=1}^{3} F_i'(x_i(T)) \, u_i(T) + F_3'(x_3(T)) \, f_3'(x_3(T))$

Now suppose that at T, the marginal fitness of tuber biomass is highest; i.e., $F_1' > F_2'$ and $F_1' > F_3'$. Then to maximize [4.1] note that for $g > 0$, $u_1(T) = 1$. Since $\lambda_i(t)$ are continuous functions of time during some time interval, there is a time t_2 such that $u_1(t) \equiv 1$ during $(t_2, T]$. Next, suppose that at t_2 one has $\lambda_2 > \lambda_1$ and $\lambda_2 > \lambda_3$. Then again, there is some time interval $(t_1, t_2]$ during which $u_2(t) \equiv 1$. Finally, suppose that at t_1 one has $\lambda_3 > \lambda_1$ and $\lambda_3 > \lambda_2$. Then there is an interval $(t_0, t_1]$ during which $u_3(t) \equiv 1$. Note that the technical details of potential discontinuities at the switching times t_i, $i = 0, 1, 2$ are ignored. Thus, the following qualitative conclusions are reached:

(a) Depending on the final marginal fitness of each tissue type, there is a final time interval during which all of the plants' photosynthesis product is invested in growth of the tissue type with the highest marginal fitness at the final time T.

(b) A switch to a growth of another tissue type may occur depending on the values of the shadow fitness (explained above); i.e., the values of the adjoint functions $\lambda_i(t)$.

An example with $t_0 = 0$ is shown in Figure 2. Note that, depending on the functional forms of g and f, more than two switches may occur. Yet, except for singular arcs (which are ignored for the sake of simplicity), the general pattern of optimal switching from a complete channelling of photosynthesis product to the growth of one tissue type to another will not change.

To proceed, let us choose, as a first approximation,

[4.4] $f_i(x_i) = a_i x_i, \quad a_i > 0 \quad \text{and} \quad g(x_3) = \alpha x_3 - \beta x_3^2$.

This means that the predation rate increases proportionately with increase in biomass of tuber, seed, and photosynthesizing tissue, with proportionality coefficients a_i. The function g includes the effect of self shading and senescence, as well as other growth limiting factors.

Suppose that the parameter values are chosen such that the growing plant exhibits only two switches, as shown in Figure 2. Then the optimal trajectory can be solved (see appendix).

To see under what circumstances there is a switch (when moving backward in time) from growth of tubers to growth of seeds, solve $\lambda_1(t) = \lambda_2(t)$ to derive t_2 (see equations [a7] and [a8] in the appendix). This gives

[4.5] $t_2 = T + \dfrac{1}{a_1 - a_2} \ln \dfrac{F_2'(x_2(T))}{F_1'(x_1(T))}$.

Therefore--as long as $\lambda_1 > \lambda_3$ and $\lambda_2 > \lambda_3$-to observe a switch from tuber growth to seed growth (in backward time) the following conditions must be satisfied:

(a) The marginal fitness of tubers at time T, $F_1'(x_1(T))$, and the marginal fitness of seeds at this time should both be either positive or negative. That is, if, for example, further increase in tuber biomass at time T results in increase in fitness, then further increase in seed biomass should do the same.

(b) If the proportionality coefficient of predation on seeds (i.e., $a_2 = (dx_2/dt)/x_2$) is greater than that of predation on tubers, that is $a_2 > a_1$, then since $t_2 < T$ one must have $|F_1'(x_1(T))| <$

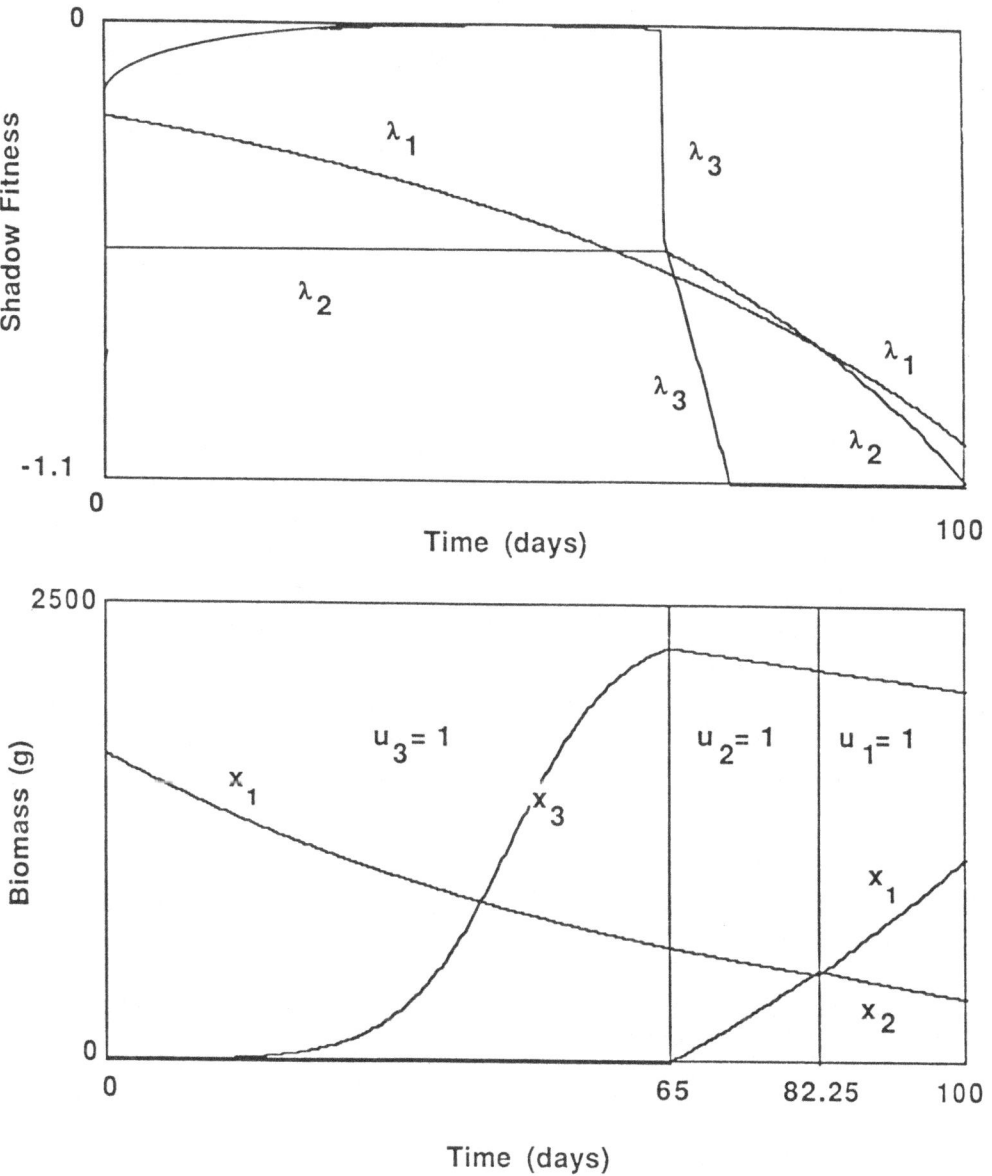

Figure 2. Optimal trajectories of the adjoint and state functions. Note how, based on the largest value of the respective adjoint variable λ_i the plant begins the 100-day growing season with growth of photosynthesizing tissue, then at day 65 it switches to growth of seeds, and at 82.25 days from the beginning of the growing season it switches to growth of tubers.

$|F_2'(x_2(T))|$. In other words, at the final time, the marginal fitness of tubers must be greater than that of seeds if both marginal fitnesses are negative, and smaller if both are positive.

(c) Since $t_2 > 0$, if $a_2 > a_1$ then [4.5] implies that

[4.6] $\qquad a_2 \geq a_2 + \dfrac{1}{T} \ln \dfrac{F_1'(x_1(T))}{F_2'(x_2(T))}$

To calculate an optimal trajectory, one has to choose, given a particular set of parameters, the final values of $x_i(T)$ and calculate the trajectories backwards such that x_{i0} are reached. This is a standard two point boundary value problem (TPBVP). Actually the problem at hand is slightly different from a standard TPBVP in that the value of $x_2(T)$ should be chosen such that $x_2(t_1) = 0$, rather than $x_2(0) = 0$. Extensive numerical simulations with various reasonable forms for $F_i(x_i(T))$ indicated that under the chosen parameter values

$a = [0.015\,,0.02\,,0.003\,]\,,\quad \alpha = 0.16,\quad \text{and}\quad \beta = 6.54 \times 10^{-5}$

F_i' must all be negative. In other words, to observe the reproductive strategy sequence of production of plant tissue, followed by production of seeds and finally tubers, the final marginal fitness of all three tissues should be negative.

A particular example with a growing season of 100 days, with parameter values as above, with the initial conditions

$x_1(0) = 1670\,,\quad x_2(0) = 0\,,\quad x_3(0) = 1.5\,,$

and with iteration at every 0.25 days is shown in Figure 2. The optimal switching times occur at $t_1 = 65$ and $t_2 = 82.25$ days from the beginning of the growing season and the final states are (in grams):

$x_1(100) = 1125\,,\quad x_2(100) = 355.6\,,\quad x_3(100) = 2039.9\ .$

The marginal fitness at the 100th day for the above values of *x* were taken as

$$F_1'(100) = -1 \, , \; F_2'(100) = -1.09 \, , \; F_3'(100) = -3.96 \, .$$

§5. Discussion

The optimal solution dictates that an optimizing growing plant should invest all of its photosynthetic product in growth of one tissue type at a time. It is an open loop solution in a sense that there is no provision for variation during the growing season. In a closed loop solution the plant should evolve some mechanisms by which the predation rates and other environmental factors which influence growth are sensed through say plant growth rate, and the assessment of growth rate is used to trigger adjustment of the optimal growth. The open loop optimal "policy" however will not change (i.e., the pattern of complete switching from growth of one tissue to another). It is conceivable that a growing plant has evolved mechanisms which allow closed loop optimization. This might be particularly true in situations where switching among growth of various tissues is not costly, and can be accomplished quickly through various metabolic pathways. Another condition which would be conducive to the evolution of feedback control is when assessment of inputs (such as environmental conditions and predation rates) forecasts, in a statistical sense, future inputs.

Most plants would exhibit suboptimal, rather than optimal growth patterns for various reasons:

(a) The evolution of optimizing mechanisms involve physiological and biochemical processes which take time and introduce delays in plant response to external inputs. The model here does not account for time delays.

(b) Switching from allocation of photosynthetic product to growth of one tissue type to another might be costly (in terms of the switch itself), and cannot be accomplished instantaneously.

(c) Competition with conspecifics will put the problem in the context of game theory, rather than the context of optimal control theory. Vincent and Brown (1984) have shown that switching from growth of photosynthesizing tissue to reproductive tissue in plants with a single reproductive strategy occur later than predicted by an optimal solution under competition in an evolutionary stable strategy game setting.

The model indicates that when the predation rates on seeds and tubers are similar, the likelihood of observing two reproductive strategies during the same season decreases; i.e., λ_1 and λ_2 become more parallel and will not intersect. The single observed reproductive strategy will then depend on which marginal fitness of the two reproductive tissues is higher. For example, in environments where predation rates on seeds and tubers are similar, but where plant density is higher, the marginal fitness of seeds by the end of the growing season might be higher, leading to phenotypes which produce seeds. The opposite may be true for phenotypes in less dense stands.

An interesting question is of course by what mechanisms does the plant "recognize" predation on various tissues. Several studies on a variety of plant species suggest that reproductive strategies may be environmentally or density controlled (Knight and Holowell 1962, Harper 1967, Jurik 1983, Bishop and Davy 1985). In some cases, plant response to grazing (e.g., shifting its reproductive effort from seed to tuber production) may be mediated through changes in density as opposed to a direct response to the actions of the herbivore. Many plant species react to changes in density by changing their reproductive strategy. For example, vegetative reproduction of *Hieracium pilosella* increases as density of plants decreased, while seed production was negatively correlated with density (Bishop and Davy 1985). Jurik (1983), in studying the sexual and vegetative reproduction of wild strawberries in different environments, suggested that vegetative reproduction is a mechanism for quickly establishing a population in an open site, while seed production is a means of dispersal from areas with a high plant density. Yeo (1965) noted that individual plants of *P. pectinatus* that produced a large number of seeds also produced fewer tubers.

An interesting extension of the above model is the case where the dynamics of the reproductive strategies operate on two time scales. The seasonal , and the long term, evolutionary time scale. When, on the evolutionary time scale, plant density affects the abundance of its herbivore, one might expect a limit cycle fluctuation of plant and herbivore densities. This means that at the beginning of some growing seasons during this limit cycle plant density is high, and a switch from a usual vegetative reproduction to reproduction through flowering and seeds may be observed. A model of this type may be invoked to explain the mysterious flowering cycle of bamboo on a 60 or 120-year cycles (Janzen 1976). The density of bamboo is highly coupled with the population dynamics of the giant panda, which depends solely on bamboo for food (Schaller 1985).

Acknowledgements. Contribution No. 15333 of the University of Minnesota Agricultural Experiment Station. Comments by Drs. Anthony Starfield and Ira Adelman helped improved the manuscript.

Literature Cited

Aiken, S.G. 1979. *North American Species of* Myriophyllum *(*Haloragaceae*).* Ph.D. Thesis, University of Minnesota.

Anderson, M.G. 1978. Distribution and production of Sago pondweed (*Potomageton pectinatus* L.) on a northern prairie marsh. *Ecology* 59: 154-160.

Anderson, M.G., and J.B. Low. 1976. Use of sago pondweed by waterfowl on the Delta Marsh, Manitoba. *The Journal of Wildlife Management* 40: 233-242.

Bellman, R. 1957. *Dynamic Programming* . Princeton University Press, Princeton, New Jersey.

Bishop, G.F., and A.J. Davy. 1985. Density and commitment of apical meristems to clonal growth and reproduction in *Hieracium pilosella. Oecologia* 66: 417-422.

Botkin, D.B., P.A. Jordan, A.S. Dominski, H.S. Lowendorf, and G.E. Hutchinson. 1973. Sodium dynamics in a northern forest ecosystem. *Proceedings of the National Academy of Sciences U.S.A.* 70: 2745-2748.

Cottam, C., 1939. *Food Habits of North American Diving Ducks.* United States Department of Agriculture Technical Bulletin Number 635.

Engel, S. 1985. *Aquatic Community Interactions of Submerged Macrophytes.* Department of Natural Resources, University of Wisconsin, Madison, Technical Bulletin Number 156.

Harper, J.L. 1967. A darwinian approach to plant ecology. *Journal of Ecology* 55: 247-270.

Isard, W., and P. Liossatos. 1979. *Spatial Dynamics and Optimal Space-time Development.* North-Holland, Amsterdam.

Iwasa, Y., and J. Roughgarden. 1984. Shoot/root balance of plants: optimal growth of a system with many vegetative organs. *Theoretical Population Biology* 25: 78-105.

Janzen, D. 1976. Why bamboos wait so long to flower. *Annual Review of Ecology and Systematics* 7: 347-391.

Jurik, T.W. 1983. Reproductive effort and CO_2 dynamics of wild strawberry populations. *Ecology* 64: 1329-1342.

King, D., and J. Roughgarden. 1982a. Multiple switches between vegetative and reproductive growth in annual plants. *Theoretical Population Biology* 21: 194-204.

King, D., and J. Roughgarden. 1982. Graded allocation between vegetative and reproductive growth for annual plants in growing season of random length. *Theoretical Population Biology* 22: 1-16.

Knight, W.E., and E.A. Holowell. 1962. Response of crimson clover to different defoliation intensities. *Crop Science* 2: 124-127.

Leitman, G. 1981. *The Calculus of Variations and Optimal Control Theory.* Plenum Press, New York.

Pontryagin, L.S., V.G. Boltyanskii, R.V. Gamkerelidze, and E.F. Mischenko. 1962. *The Mathematical Theory of Optimal Processes.* Wiley, New York.

Schaller, G.B. 1985. *The Giant Pandas of Wolong.* University of Chicago Press, Chicago.

Vincent, T.L. 1979. Yield for annual plants as an adaptive response. *Rocky Mountain Journal of Mathematics* 9: 163-173.

Vincent, T.L., and H.R. Pulliam. 1980. Evolution of life history strategies for an asexual annual plant model. *Theoretical Population Biology* 17: 215-231.

Vincent, T.L., and J.S. Brown. 1984. Stability in an evolutionary game. *Theoretical Population Biology* 26: 408-427.

Yeo, R.R. 1965. Life history of sago pondweed. *Weeds* 13: 314-321.

Appendix: Solutions for Optimal Trajectory with Two Switches

Suppose that by the end of the growing season the marginal fitness due to the final biomass of tubers is greater than that of both seeds and photosynthesizing tissue. Then λ_1 is larger than both λ_2 and λ_3 during some time interval $[t_2, T]$. Now suppose that at t_2, λ_2 becomes larger (as one moves backwards in time) than both λ_1 and λ_3 and remains so until $t_1 < t_2$. Finally, suppose that at t_1, λ_3 becomes larger than both λ_1 and λ_2 and remains so until $t = 0$. Then the plant should first grow; i.e., $u_3(t) \equiv 1$ for $t \in [0, t_1]$. At t_1, $u_2(t) = 1$ and remains so until t_2, and therefore all of the photosynthesis product is invested in seed growth. Finally, between t_2 and T the optimal solution stipulates that $u_1(t) \equiv 1$ and all of the photosynthesis product is invested in tuber growth. With the appropriate values of u_i substituted in equations [2.1], [2.2], [2.5], and [2.6], the exact solution of the optimal trajectory becomes as follows.

(I) $t \in [t_2, T]$

During this time interval, since $u_1 \equiv 1$, equations [2.1], [2.2], [2.5], and [2.6] become

[a1] $dx_1 / dt = (\alpha - \beta x_3) x_3 - a_1 x_1 , \qquad \lambda_1 = a_1 \lambda_1$

[a2] $dx_2 / dt = -a_2 x_2, \qquad \lambda_2 = a_2 \lambda_2$

[a3] $dx_3 / dt = -a_3 x_3, \qquad \lambda_3 = -(\alpha - 2\beta x_3) \lambda_1 + a_3 \lambda_3 .$

Solving backwards in time, where the remaining time is denoted by $\tau = T - t$, gives

[a4] $x_1(t) = x_1(T)\, e^{a_1\tau} - \dfrac{\alpha x_3(T)}{a_1 - a_3} \left[e^{a_1\tau} - e^{a_3\tau} \right] + \dfrac{\beta [x_3(T)]^2}{a_1 - 2a_3} \left[e^{a_1\tau} - e^{2a_3\tau} \right]$

[a5] $x_2(t) \doteq x_2(T)\, e^{a_2\tau}$

[a6] $x_3(t) = x_3(T)\, e^{a_3\tau}$

[a7] $\lambda_1(t) = F_1{}'(x_1(T))\, e^{-a_1\tau}$

[a8] $\lambda_2(t) = F_2'(x_2(T)) e^{-a_2\tau}$

[a9] $\lambda_3(t) = F_3'(x_3(T)) e^{-a_3\tau} + \dfrac{\alpha F_1'(x_1(T))}{a_1 - a_3} \left[e^{-a_3\tau} - e^{-a_1\tau} \right]$

$$- \dfrac{2\beta x_3(T) F_1'(x_1(T))}{a_1 - 2a_3} \left[e^{-a_3\tau} - e^{-(a_1 - a_3)\tau} \right]$$

(II) $t \in [t_1, t_2]$

During this time interval $u_2 \equiv 1$ and equations [2.1], [2.2], [2.5], and [2.6] become

[a10] $dx_1 / dt = - a_1 x_1,$ $\qquad\qquad \lambda_1 = a_1 \lambda_1$

[a11] $dx_2 / dt = (\alpha - \beta x_3) x_3 - a_2 x_2,$ $\qquad \lambda_2 = a_2 \lambda_2$

[a12] $dx_3 / dt = -a_3 x_3,$ $\qquad\qquad \lambda_3 = -(\alpha - 2\beta x_3) \lambda_2 + a_3 \lambda_3.$

Solving backwards in time, where the remaining time is now denoted by $\tau = t_2 - t$, gives

[a13] $x_1(t) = x_1(t_2) e^{a_1\tau}$

[a14] $x_2(t) = x_2(t_2) e^{a_2\tau} - \dfrac{\alpha x_3(t_2)}{a_2 - a_3} \left[e^{a_2\tau} - e^{a_3\tau} \right] + \dfrac{\beta x_3(t_2)^2}{a_2 - 2a_3} \left[e^{a_2\tau} - e^{2a_3\tau} \right]$

[a15] $x_3(t) = x_3(t_2) e^{a_3\tau}$

[a16] $\lambda_1(t) = \lambda_1(t_2) e^{-a_1\tau}$

[a17] $\lambda_2(t) = \lambda_2(t_2) e^{-a_2\tau}$

[a18] $\lambda_3(t) = \lambda_3(t_2) e^{-a_3\tau} + \dfrac{\alpha \lambda_2(t_2)}{a_2 - a_3} \left[e^{-a_3\tau} - e^{-a_2\tau} \right]$

$$- \dfrac{2\beta x_3(t_2) \lambda_2(t_2)}{a_2 - 2a_3} \left[e^{-a_3\tau} - e^{-(a_2 - a_3)\tau} \right]$$

(III) $t \in [0, t_1]$

After substituting $u_1 \equiv u_2 \equiv 0$ and $u_3 \equiv 1$ during this time interval, the solution, with $\tau = t_1 - t$, becomes

[a19] $x_1(t) = x_1(t_1) \, e^{a_1 \tau}$

[a20] $x_2(t) \equiv 0$

[a21] $x_3(t) = \dfrac{x_3(t_1) \, (\alpha - a_3)}{\beta x_3(t_1) + [\alpha - a_3 - \beta x_3(t_1)] \, e^{(\alpha - a_3)\tau}}$

[a22] $\lambda_1(t) = \lambda_1(t_1) \, e^{-a_1 \tau}$

[a23] $\lambda_2(t) \equiv \lambda_2(t_2)$

For $A = \beta x_3(t_1)$, $B = a_3 - \alpha$, and $C = -[A+B] \, e^{-B\tau}$ we get

[a24] $\lambda_3(t) = \lambda_3(t_1) \exp \{ B\tau - 2 \ln |A + Ce^{Bt_1}| + 2 \ln |A + Ce^{Bt}| \}$

APPLICATIONS OF OPTIMAL IMPULSE CONTROL
TO OPTIMAL FORAGING PROBLEMS

Yosef Cohen

Department of Fisheries and Wildlife
University of Minnesota, St. Paul, Minnesota, 55108 USA

Abstract. The method of optimal impulse control is applied to problems of optimal foraging and optimal switching between patches. This method is particularly appropriate for analysis and modeling of many biological processes during which there is a sudden switch from a particular level of the states to another. It allows derivation of the optimal continuous control and of the optimal time at which a switch from one level of states to another should occur. Three models are considered, two with optimal foraging on a non-renewable patchy resource, and one with renewable patchy resource. Optimal switching times between patches are derived explicitly for the former two and discussed for the latter.

§1. Introduction

Many biological processes involve a sudden change from one level of states to another. Examples are the digestive process, in which the states of food in the digestive tract (where states define quantities of interest, such as concentration of energy, or other nutrients) change abruptly as food is swallowed; an animal foraging in a patchy habitat where the time of switching between patches is short relative the the food dynamics in a patch; changes in concentration of hormones or other substances in the blood; change in the proportion of photosynthesis products allocated to vegetative growth or growth of reproductive tissue in plants. From an optimal control perspective, there are two problems involved in such processes: first, how to choose controls, and second how to choose switching times such that a particular criterion is maximized (or minimized). These then are the substance of impulse control (Blaquiere 1979), which has many potential applications in evolutionary biology and ecology.

My purpose is to introduce the method and demonstrate by way of examples how it may be used to derive optimal switching times from one level of states to another when the switch itself involves a cost. The results from optimal control theory are quoted-- with significant simplification--from Blaquiere (1979) in the appendices. Examples which suggest possible applications are discussed, beginning with a simple problem.

§2. Applications to Non-renewable Resources

The following sections demonstrate the use of Theorem 1 in Appendix 2, and its potential applications to optimal foraging problems. This section deals with optimal foraging on non-renewable resources, and in section §3 I discuss (briefly) the case of optimal foraging on a renewable resource. Note that with slight changes of definitions of states, etc., the problem can be applied to other models; e.g., fisheries models in which the fisher considers when to switch from one fishing ground to another to maximize harvest. The model in §2.1 is simple, and as discussed at the end of this section can be solved directly rather easily. For heuristic reasons it is discussed in some detail.

§2.1. Example: Optimal Switching Between Patches

Consider a forager feeding during a time interval $[0, T]$. It exhausts its resource at a rate that is proportional to the amount of resource left; i.e., $dx/dt = -u\,x(t)$ where $x(t)$ is the amount of resource left in the patch and $u \in [0, U]$. In other words, the forager controls the rate at which it forages, and due to physical limitations this rate cannot exceed U. Now suppose that an animal is observed during the specified time interval to switch between two patches only, and suppose the animal estimates the initial density of food in patch zero and patch one, x_0 and x_1, respectively. Then according to the notation in Appendix 1 the model is

[2.1] $dx/dt = -u(t)\,x(t) + \mu(t)\,[x_1 - x(t)]$, $\quad x(0) = x_0$, $\quad t \in [0, T]$, where $x(t)$ denotes the amount of food remaining in the patch currently being visited at time t, $u \in [0, U]$ is the foraging intensity control variable, $\mu \in [0, 1]$ is the switching control variable (i.e., when $\mu = 0$ no switching occurs and when $\mu = 1$ a switch to the next patch occurs), and $[0, T]$ is the interval over which foraging is observed.

Since the rate of food intake equals the rate at which food is exhausted from the habitat, then assume that maximizing the net intake rate during $[0, T]$ maximizes fittness. Let the cost of switching between the two patches be C, and the net return from a unit of forage be R. Then according to the notation in Appendix 1 the objective functional to be minimized is

[2.2]
$$V(x^{o}, s, x(\bullet)) = \int_{0}^{T} - R\, u\, x(t)\, dt + C\, \mu(t_{c})$$

where t_c denotes the switching time. According to the notation introduced in Appendix 1, the problem can be separated into two parts: the continuous control problem (denoted by f_i, $i = 0, 1$ and u), and the impulse control problem (denoted by g_i and μ):

[2.3]
$$f_0(x, u) = -Rxu, \quad g_0(x, \mu) = \mu C, \quad f_1(x, u) = -ux, \quad g_1(x, \mu) = \mu(x_1 - x),$$

$$0 \leq u \leq U, \quad 0 \leq \mu \leq 1, \quad \lambda_0(t) \equiv 1, \quad x > 0, \quad R > 0, \quad C > 0.$$

where λ denotes the adjoint variable. With H denoting the "continuous" Hamiltonian and H_c denoting the "impulse" Hamiltonian (see Appendix 2):

[2.4] $\qquad H(\lambda, x, u) = -u\, x\, (R + \lambda)$

[2.5] $\qquad H_c(x, \mu) = \mu\, [\, C + \lambda(t_c+0)\, (x_1 - x)\,]$

[2.6] $\qquad d\lambda / dt = R\, u + \lambda\, u, \quad \lambda(T) = 0$

[2.7] $\qquad \lambda(t_c) = \lambda(t_c+0) - \mu\, \lambda(t_c+0)\, .$

where $\lambda(t_c+0)$ denotes the instant of time just after switch occured at time t_c. At $t = T$, $\lambda(t) = 0$. Therefore from [2.4] and condition (i) of Appendix 2, $u = U$. Also at T, $d\lambda/dt = Ru$. Therefore λ is increasing at T. Thus, there is some interval for which $R > \lambda$. During this interval $u = U$ and u will not switch to 0 as long as $R + \lambda > 0$. Solving [2.6] backwards in time gives

[2.8] $\qquad \lambda(t) = -R[1 - e^{U(T-t)}], \quad t \in (t_c, T]$.

Since $R + \lambda > 0$ for all $t \in (t_c, T]$, $u = U$ throughout this interval. From [2.8]

[2.9] $\qquad \lambda(t_c+0) = -R\left[1 - e^{-U(T-t_c)}\right]$.

From condition (ii) in Appendix 2 and [2.5] with [2.9] one gets

$$
\begin{array}{llll}
& a & \mu = 1 & \text{if} \quad C - R\left[1 - e^{-U(T-t_c)}\right]\left[x_1 - x\right] < 0 \\
[2.10] & b & \mu = 0 & \text{if} \quad C - R\left[1 - e^{-U(T-t_c)}\right]\left[x_1 - x\right] > 0 \\
& c & \mu \text{ undefined} & \text{if} \quad C - R\left[1 - e^{-U(T-t_c)}\right]\left[x_1 - x\right] = 0 .
\end{array}
$$

Note that μ denotes the switching variable (0 or 1), and t_c denotes the switching time. [2.10c] defines the boundary between [2.10a] and [2.10b], and [2.10b] defines the region in which no switch can occur. To delineate these regions solve [2.10c] for $\phi(t_c) = x(t_c)$ to get

[2.11] $\qquad \phi(t_c) = \dfrac{x_1 R\left[1 - e^{-U(T-t_c)}\right] - C}{R\left[1 - e^{-U(T-t_c)}\right]}$

Note that $\phi(t_c)$ is a decreasing and concave function of t_c. Therefore, for $t_c > 0$ the requirement that $\phi(0) > 0$ must be imposed. This leads to the following restrictions on the parameters:

[2.12]   (a) $x_1R > C$ and (b) $UT > \ln \dfrac{x_1R}{x_1R - C}$

In other words, for a switch to occur, the total expected net return from the next patch must exceed the cost of switching, and UT must be large enough, so that [2.12b] is satisfied. If condition (a) is not satisfied then [2.10b] will be true throughout $[0, T]$.

From [2.7]

[2.13] $\lambda(t_c) = \lambda(t_c+0) - \mu \lambda(t_c+0) = 0$.

Thus, the same arguments leading to $u = U$ for $t \in (t_c, T]$ can be used to derive $u = U$ for $t \in [0, t_c)$. This result is also intuitively clear from the objective [2.2].

[2.13] with condition (iii) of Appendix 2 results in

[2.14] $x(t_c) = x_1 e^{-U(T-t_c)}$.

Define $\psi(t_c) = x(t_c)$. Figure 1 shows ϕ and ψ. A switch can occur along the arc AB only. The optimal switching time t_c^* is found by equating $x(t) = x_0 e^{-Ut}$ to [2.14] and solving for t_c^*, which results in

[2.15] $t_c^* = \dfrac{T}{2} + \dfrac{1}{2U} \ln \dfrac{x_0}{x_1}$.

The effect of the various parameters on the size of the region of no switch and on t_c^* can be summarized from [2.11], [2.14], and [2.15]:

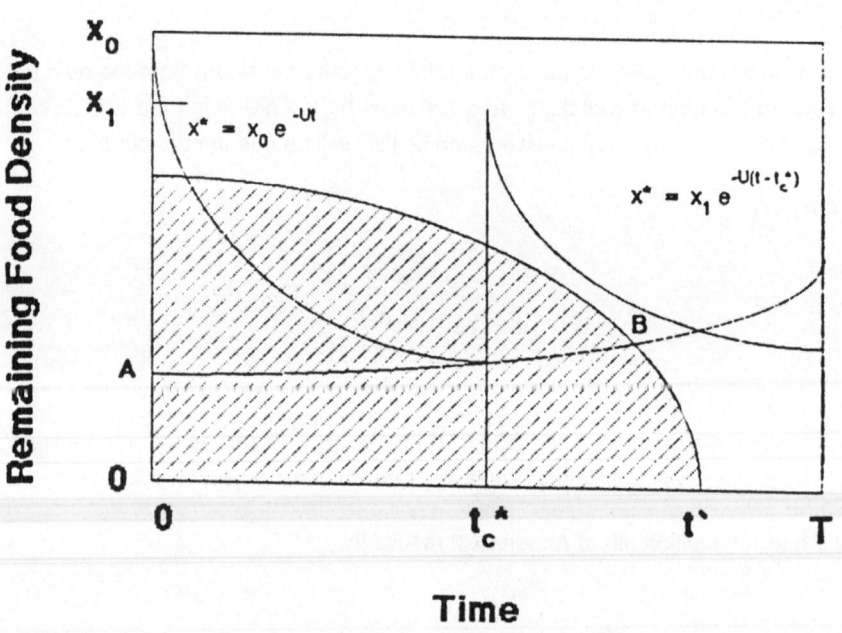

Figure 1. The shaded region is bounded by $\phi(t_c)$ and the arc $\psi(t_c)$ within it (AB) is the region in which a switch will occur. The optimal switch occurs when the optimal trajectory $x^*(t) = \psi(t_c)$, shown as t_c^*.

(a) As R or x_1 increase, the boundary between region of potentially occuring switch (shaded area in Figure 1) and region of no switch will shift to the right, and the length of the arc AB along which switching can occur (equation [2.14]) will increase. In other words, the region of switching increases (see Figure 1).

(b) As U increases, the trajectory for $x(t)$ starting at $t = 0$ will be more likely to hit $\psi(t_c)$, and the likelihood of observing a switch increases.

(c) As the ratio x_0/x_1 decreases, switching will occur earlier.

(d) Increase in T will shift $\phi(t_c)$ to the right, and thus increase the likelihood of switching.

Note that the problem solved in this section can be solved rather easily--with some minor difference--by maximizing

$$f(t_c) = x_0 \left[1 - e^{-Ut_c} \right] + x_1 \left[1 - e^{-U(T-t_c)} \right] - C$$

with respect to t_c. This resembles the static approach taken by Charnov (1976) in his development of the marginal value theorem. Yet, as demonstrated in the following sections, the approach taken here allows for considerable flexibility in choosing models.

§2.2. Example: Optimal Switching Between Patches; Continued

Next, assume that the model in the previous example holds, except that now the cost of switching is a function of how much the animal has consumed. In other words, assume C is now $C(x)$. Since x is the amount of food remaining, the less food remains, the higher C is. Therefore $C(x)$ is a decreasing function of x. In particular, consider as a first approximation the function

[2.16] $C(x) = C` - \beta x.$

In [2.16] C is the maximum cost of switching, which occurs when no resource remains in the habitat, and β quantifies the rate at which the switching cost increases when x decreases.

According to the notation in Appendix 1 one gets:

[2.3'] $f_0(x, u) = -R x u$, $g_0(x, \mu) = \mu (C - \beta x)$, $f_1(x, u) = -u x$,
$g_1(x, \mu) = \mu (x_1 - x)$, $0 \leq u \leq U$, $0 \leq \mu \leq 1$, $\lambda_0(t) \equiv 1$, $x > 0$,
$C > 0$, $\beta > 0$, $R > 0$.

[2.4] remains unchanged and [2.5] becomes

[2.5'] $H_c(x, \mu) = \mu [C - \beta x + \lambda(t_c+0) (x_1 - x)]$.

[2.6] remains unchanged and [2.7] becomes

[2.7'] $\lambda(t_c) = \lambda(t_c+0) - \mu [\beta + \lambda(t_c+0)]$.

[2.8] remains unchanged. Just after switching [2.9] holds, and from condition (ii) of Appendix 2 and [2.5']

        a' $\mu = 1$ if $C - \beta x + \lambda(t_c+0) [x_1 - x] < 0$

[2.10] b' $\mu = 0$ if $C - \beta x + \lambda(t_c+0) [x_1 - x] > 0$

        c' μ undefined   if $C - \beta x + \lambda(t_c+0) [x_1 - x] = 0$

Thus,

[2.11'] $\phi(t_c) = \dfrac{R \left[1 - e^{-U(T-t_c)} \right] - C}{R \left[1 - e^{-U(T-t_c)} \right] - \beta}$.

For $\phi(t_c) > 0$ to occur both the numerator and denominator of [2.11'] must be positive or negative. To simplify the analysis, assume that the cost of switching to the next patch becomes very high when the resource is almost exhausted. In this case the numerator in [2.11'] is negative and for $\phi(t_c)$ to be positive the numerator must be negative as well. In this case, the following restrictions on the parameters will ensure switching during $[0, T]$:

$$x_1 R > C`, \text{ and } R > \beta.$$

Under the condition of negative numerator and denominator in [2.11'] one may conclude that:

(a) As $C`$ increases the boundary shifts to the left, and the region of no switch increases.

(b) Increase in β shifts the boundary to the right, and the arc along which a switch can occur becomes longer.

(c) x_1, R and U have the same effect on the boundary as in [2.11].

To calculate $t_c{}^*$ note that from [2.7'] and Theorem 1 (in Appendix 2)

[2.13'] $\lambda(t_c) = -\beta$

and from [2.9] with condition (iii) of Appendix 2

[2.14'] $\psi(t_c) = \dfrac{R x_1 e^{-U(T-t_c)}}{R - \beta}$

The optimal switching time is found by equating [2.14'] to $x_0 e^{-Ut}$ to give

[2.15'] $$t_c^* = \frac{T}{2} + \frac{1}{2U} \ln \frac{x_0 (R - \beta)}{x_1 R} .$$

Note that when $\beta = 0$, [2.14'] and [2.15'] reduce to [2.14] and [2.15]. As β increases, the optimal switching time between patches, t_c^*, becomes smaller; i.e., when the cost of switching increases more rapidly with more food consumed (and therefore with less food remaining in the environment), then to behave optimally, the forager switches earlier.

§3. Example: Optimal Switching Between Patches with Renewable Resource

Suppose now that the resource is renewable, and there are two patches. With two patches with known initial food density, and with u denoting the harvest rate of x, the model becomes

[3.1] $$dx/dt = f(x) - u + \mu (x_1 - x) , \quad x(0) = x_0 .$$

Suppose now that the return from a unit of x is a constant R, and the cost of obtaining a unit of x is $S(x)$. Denote the cost of switching between patches by C. The forager is maximizing its total net intake during $[0 , T]$. The objective is then to minimize

[3.2] $$\int_0^T -[R - S(x)] u(t) \, dt + \mu C$$

with u and x restricted as before and $dS/dx < 0$ with $S(x) > 0$ (see Clark 1976).

The Hamiltonians become

[3.3] $\qquad$ $H = -[R - S(x)]u + \lambda(t)[f(x) - u]$

[3.4] $\qquad$ $H_c = \mu\{C + \lambda(t_c+0)[x_1 - x(t_c)]\}$.

According to condition (i) of Theorem 1, H is minimized when

$\qquad$ a $\qquad$ $u = U$ $\qquad$ if $S(x) - R - \lambda(t) < 0$

[3.5] $\qquad$ b $\qquad$ $u = 0$ $\qquad$ if $S(x) - R - \lambda(t) > 0$

$\qquad$ c $\qquad$ u singular $\quad$ if $S(x) - R - \lambda(t) = 0$.

Also

[3.6] $\qquad$ $d\lambda/dt = -S'(x)u - \lambda(t)f'(x)$

[3.7] $\qquad$ $\lambda(t_c) = \lambda(t_c+0) - \lambda(t_c+0)\mu$

where primes denote derivatives with respect to the dependent variable. Ignoring the singular control for the moment, note that since $\lambda(T) = 0$, then: (a) $u = U$ if $S(x(T)) < R$; (b) $u = 0$ if $S(x(T)) > R$; and (c) u is at its singular value if $S(x(T)) = R$. When (b) occurs, $u = 0$ and $d\lambda(T)/dt = 0$. This means that $\lambda(t) = 0$ for $t \in [0, T]$, and the switching to U depends on $S(x(t)) - R$.

If the animal is actually observed foraging at T, then $S(x(T)) < R$ (ignoring the singular case for the moment). Then at T, $u = U$, and $d\lambda(T)/dt = -S'(x(T))U$. Since $S(x(t)) > 0$ for all t, then at the final time $\lambda(T)$ is decreasing to zero. In other words, $\lambda(t)$ is increasing in backwards time from zero, and there is some time interval over which $u = U$. The switch to singular control will occur when $\lambda(t_s) = S(x(t_s)) - R$. At this time the control will become singular, and will continue in backwards time to be singular until a time at which it will switch again to either 0 or U, so that $x(0) = x_0$. This solution was discussed in more detail by Clark (1976). If $x(T) = x^*$, where x^* is the optimal value of

x at the singular control, then only one switch will occur at a time that will bring x to x_0 by using either $u = 0$ or $u = U$.

For singular control one gets

[3.8] $S'(x)f(x) + [S(x) - R]f'(x) = 0$

(see Clark 1976). Therefore, the singular control should be such that $d[S(x)f(x)]/dx = Rf'(x)$.

To proceed, choose $f(x) = \alpha x - \beta x^2$ and $S(x) = \gamma/x$ (see Schaefer 1957 and Clark 1976). Then from [3.8]

[3.9] $(-\gamma/x^2) \times (\alpha - \beta x) + [\gamma/x - R][\alpha - 2\beta x] = 0$

which gives the optimal value of x for the singular control

[3.10] $x^* = (R\alpha + \beta\gamma) / 2R\beta.$

From [3.1] with $dx/dt = 0$ the optimal value of u during singular control is

[3.11] $u^* = [\alpha(R\alpha + \beta\gamma) - \beta(R\alpha + \beta\gamma)^2]/2R\beta.$

Now suppose that the animal is observed to forage along the singular trajectory, then it switches to $u = U$ at time t_s, and then, before $t = T$ it switches to the new patch at time t_c. Just after switching at t_c, $u = U$ if $x_1 > x^*$. The solution for t_c^* and t_s^* then proceeds numerically as in the previous sections.

§4. Discussion

The results derived in §2.1 are much like the marginal value theorem of Charnov (1976), where the cost of switching was implicitly considered in terms of the travel time. In §2.2 the cost of switching becomes a function of both the cost of switching between patches and of the condition of the animal (in terms of the amount of food it has already consumed). The problem in §2.1 was discussed more thoroughly in Case (1975) in the context of optimal repainting of a roadside inn.

The derivations here pertain to switching between two patches only. The extension to more than two patches could be accomplished algorithmically by calculating t_c^* for each pair of patches moving backward in time. It should be pointed out that the problem in §2.1 for two patches only can be solved by simple calculus. A different approach to derivation of optimal switching time was proposed by Koopman (1980) in the context of search theory (Stone 1975), and discussed in detail by Mangel (1985) in the context of optimal search for schools of fish. An interesting extension of the problem arises when x_i, the initial food density in the next patch is not known with certainty, but its pdf is. Such extensions may be applied with the techniques discussed in Mangel (1985).

Acknowledgments. Contribution number 15297 of the University of Minnesota Agricultural Experiment Station. I thank Marc Mangel for pointing out to me the fact that the solution of the model in §2.1 can be derived by simple calculus and bringing Koopman's algorithm to my attention.

Literature Cited

Blaquiere, A. 1977. Differential games with piece-wise continuous trajectories. In P. Hadedorn, H.W. Knobloch, and G.J. Olsder (editors). *Lecture Notes in Control and Information Sciences*, Volume 3, Springer Verlag, Berlin.

Blaquiere, A. 1979. Necessary and sufficiency conditions for optimal strategies in impulsive control. In P. Liu and E. Roxin (editors). *Differential Games and Control Theory III*. Marcel Kekker, New York.

Case, J. 1975. Impulsively controlled differential games. In J.D. Grote (editor). *The Theory and Application of Differential Games*. D. Reidel, Dordrecht.

Charnov, E.L. 1976. Optimal foraging, the marginal value theorem. *Theoretical Population Biology* 9: 129-136.

Clark, C.W. 1976. *Mathematical Bioeconomics: The Optimal Management of Renewable Resources*. Wiley, New York.

Leitman, G. 1981. *The Calculus of Variation and Optimal Control*. Plenum Press, New York.

Mangel, M. 1985. *Decision and Control in Uncertain Resource Systems*. Academic Press, New York.

Schaefer, M.B. 1957. Some considerations of population dynamics and economics in relation to the management of marine fisheries. *Journal of the Fisheries Research Board of Canada* 14: 669-681.

Stone, L.D. 1975. *Theory of Optimal Search*. Academic Press, Yew York.

§Appendix 1. Problem Statement

The problem statement and notation are introduced formally in this section. This and the next section are taken almost verbatim from Blaquiere (1979), and are cited here for ease of reference. Although difficult at first, I found the notation and problem statement in this format very useful for later applications. Technical details are overlooked for clarity, and the interested reader should consult Blaquiere (1979).

Consider the state $x(t) = [x_1(t), \ldots, x_n(t)]$ where $x_n(t) = t$, and $x \in X$ R^n. Let u and μ denote the regular and impulse control variables. Let K_u and K_μ be prescribed non-empty subsets of $U \in R^{d_1}$ and of $M \in R^{d_2}$, respectively. Let $f(\cdot)$ and $g(\cdot)$ be prescribed functions of class C^1 such that

$$f(\cdot): X \times U \to R^n \qquad (x, u) \to f(x, u)$$

$$g(\cdot): X \times M \to R^n \qquad (x, \mu) \to g(x, \mu)$$

where $f(\cdot) = [f_1, \ldots, f_n]$, $g(\cdot) = [g_1, \ldots, g_n]$, $f_n(x, u) \equiv 1$, $g_n(x, \mu) \equiv 0$. Consider a "player" J_0 who controls $x(t)$ through a choice of strategy $s \in S_0$, where S_0 is a prescribed strategy set. The following definitions will fix the notation for later use.

Definition 1. A strategy $s = (Y, p(\cdot), \pi(\cdot)) \in S_0$ is admissible if and only if $x \in Y \Rightarrow x + g(x, \pi(x)) \in X - Y$.

Let t-0 and t+0 denote the instant of time just before and just after switching occurs, respectively.

Definition 2. $x(t)$, $t \in [t_0, t_f]$, is a path in R^n generated by $s \in S$ (where S is the set of all admissible strategies) from an initial state x^0 if and only if:

1) $x(t_0) = x^0$;

2) $x(t)$ is piecewise continuous on $[t_0, t_f]$;

3) $x(t) = x(t$-$0)$ for $t \neq t_0$;

4) $t \in T[t_0, t_f] \Rightarrow x(t) \in Y$ and $x(t+0) = x(t) + g(x(t), \pi(x(t)))$ where $T[t_0, t_f]$ denotes the set of discontinuity points of $x(t)$;

5) for all $t \notin T[t_0, t_f]$, except possibly t_f, $x(t) \in X - Y$;

6) $dx(t)/dt = f(x(t), p(x(t)))$ for all $t \in [t_0, t_f]$, except on at most a denumerable subset of $[t_0, t_f]$.

Let θ denote the target set. J_0 desires to steer $x(t)$ from x^0 to $x(t_f) \in \theta$.

Definition 3. $x(t)$, $t \in [t_0, t_f]$, generated by $s \in S$ from x^0 is a terminating path if and only if $x(t_f) \in \theta$.

Definition 4. $s \in S_0$ is playable at x^0 if and only if it is: (a) admissible and (b) it generates a terminating path from x^0.

Let

$$f_0(\cdot): X \times U \to R \qquad (x, u) \to f_0(x, u)$$

$$g_0(\cdot): X \times M \to R \qquad (x, \mu) \to g_0(x, \mu).$$

Let $\theta_0(x(t_f)) \in \theta$ be a function of class C^1. The cost of path $x(t)$, $t \in [t_0, t_f]$, generated by $s = (Y, p(\cdot), \pi(\cdot)) \in S$ from x^0 is

[A1.1]
$$V(x^0, s, x(\bullet)) = \theta_0(x(t_f)) + \int_{t_0}^{t_f} f_0(x(t), p(x(t))) \, dt + \sum_{t \in T[t_0 t_f]} g_0(x(t), \pi(x(t))).$$

J_0 desires to minimize V. Denoting by $J(x^0)$ the set of all playable strategies at x^0 and by $I(x^0, s)$ the set of all terminating paths, then

Definition 5. s^* is optimal at x^0 if and only if: (a) s^* is playable at x^0, and (b) $V(x^0, s^*, x^*(\cdot)) \leq V(x^0, s, x(\cdot))$ for $s \in J(x^0)$, all $x(\cdot) \in I(x^0, s)$, and all $x^*(\cdot) \in I(x^0, s^*)$.

§Appendix 2. Necessary Conditions for Optimal Strategy

The statement of necessary conditions is preceded in Blaquiere (1979) by a set of assumptions which assure that f and g are defined during the jump process. These assumptions will be met by most models which describe biological processes, and will not be cited. Readers interested in applying the technique should consult Blaquiere (1979).

Let

[A2.1] $\lambda(t)$: $[t_i , t_j] \rightarrow R^{n+1}$ be piecewise continuous with $\lambda(t) = \lambda(t-0)$ for $t \in (t_i , t_j]$,

[A2.2] $H(\lambda , x , u) = \sum_{\alpha=0}^{n-1} \lambda_\alpha f_\alpha(x , u) ,$ $\lambda \equiv \lambda(t) ,$ and

[A2.3] $H_c(x , \mu) = \sum_{\alpha=0}^{n-1} \lambda_\alpha(t_c+0) g_\alpha(x , \mu) .$

Theorem 1. There exists $\tau > 0$ and $\lambda_\alpha(\bullet)$, $\alpha = 0 , 1 , ... , n\text{-}1$ solutions of

[A2.4] $\dfrac{d\lambda_\alpha}{dt} = - \dfrac{\partial H}{\partial x_\alpha} \Big|_{x=x^*(t) , u=p^*(x)}$ with

[A2.5] $\lambda_\alpha(t_c) = \lambda_\alpha(t_c+0) + \dfrac{\partial H_c}{\partial x_\alpha} \Big|_{x=x^*(t_c) , \mu=\pi^*(x)}$

on $[t_c\text{-}\tau , t_c)$ and $(t_c , t_c+\tau]$ such that

(i) $\min_{u \in K_u} H(\lambda(t) , x^*(t) , u) = H(\lambda(\tau) , x^*(t) , p^*(x^*(t)))$

for all $t \in [t_c\text{-}\tau , t_c+\tau]$, $t \neq t_c$.

(ii) $\min_{\mu \in K_\mu} H_c(x^*(t_c) , \mu) = H_c(x^*(t_c) , \pi*(x^*(t_c)))$

(iii) $\min\limits_{u \in K_u} H(\lambda(t_c+0), x^*(t_c+0), u) - \min\limits_{u \in K_u} H(\lambda(t_c), x^*(t_c), u)$

$$= \frac{\partial H_c(x, \mu)}{\partial x_\alpha} \Bigg|_{x=x^*(t_c), \mu=\pi^*(x)} .$$

(iv) $\lambda_0(t) \equiv 1$ for all $t \in [t_c-\tau, t_c+\tau]$.

(v) $\min\limits_{u \in K_u} H(\lambda(t_c+0), x^*(t_c+0), u) - \min\limits_{u \in K_u} H(\lambda(t_c), x^*(t_c), u) \leq$

$$= \frac{\partial H_c(x, \mu)}{\partial x_\alpha} \Bigg|_{x=x^*(t_c), \mu=\pi^*(x)} .$$

When $x_\alpha = t_f$ then the usual transversality condition (Lietman 1981) becomes

$$\lambda_\alpha(t_f) - \frac{\partial \theta_0}{\partial x_\alpha} \Bigg|_{x=x^*(t_f)} , \qquad u = 1, \ldots, n-1 .$$

The proof of Theorem 1 may be found in Blaquiere (1977).

FUR SEAL AND BLUE WHALE:
THE BIOECONOMICS OF EXTINCTION

Robert McKelvey

Department of Mathematical Sciences
University of Montana, Missoula, Montana 59812 USA

Abstract. Common property exploitation repeatedly has been implicated in instances of mismanagement of marine biological resources, resulting in depletion, even exhaustion, of stocks, and in dissipation of the economic benefits that the harvest might entail. Here I shall re-examine historical patterns in the competitive exploitation of two marine mammals, examining the interplay between common property harvest practices and inertial effects that result, among other things, from irreversible capital investment ("sunk capital") in the harvesting industry. I find that common property exploitation tends to exaggerate the swings and overshoots that inertial features introduce into the temporal pattern of harvesting, leading to an excessive buildup of capital capacity, followed by an excessive depletion of the resource stock. Under some conditions these exaggerated swings can result in stock extinctions which optimal management might have avoided. The formal model is set up as a differential game, and analyzed by control theory methods.

§1. Bioeconomic Predator-Prey Models

The obvious parallel, between predator-prey interactions in natural ecosystems and competitive harvesting of a wild biological resource, has inspired considerable theoretical analysis and model-building.

In his seminal article, "The economic theory of a common-property resource: The fishery," Gordon (1954) argued that, in harvesting a previously unexploited fish stock, "man's intrusion would have the effect of any other predator; and that can only mean that the species population would reach a new equilibrium at a lower level of abundance." More decisively, Gordon enunciated the principle that the human predator population size, governed by the interaction between biological and economic forces, would come into a certain "bioeconomic equilibrium" with the prey population. This would occur at a level of harvest effort which, having driven down the reproducing stock of fish, would yield a rate of income from harvest only just balancing the costs (including "opportunity costs") of exerting the effort.

A full predator-prey dynamic model of this process was proposed by Smith (1969). It incorporates the standard "surplus production" harvest model for the prey dynamics

[1.1] $dx/dt = F(x) - h(t)$,

with $x(t)$ being stock size, $F(x)$ intrinsic growth rate, and $h(t)$ harvest rate. Harvest is assumed to be proportional to both stock size and fishing effort level; thus $h = qxK$, where $K(t)$ is a measure of the size of the active fishing fleet.

Smith's model couples the harvest equation [1.1] with a second equation which describes the evolution in the size of the fleet. The simple behavioral rule is that the rate of entry of vessels into the fleet (or of exit from the fleet) is proportional to the current per-unit net profit (or loss) rate

$\pi/K = pqx - c$.

Here π is the net profit, p is the unit price of harvest, and c is the unit vessel cost. Hence

[1.2] $dK/dt = \beta [pqx - c]$,

with β an empirical proportionality constant.

A voluminous theoretical literature has developed around this dynamic predator-prey formulation and its variants. (For a partial account, see Berck 1979). From the general structure of the equations, it is clear that--assuming downward-sloping per-capita growth rate--there is a locally-stable equilibrium at Gordon's bioeconomic *point* $x_\infty = c / pq$, $K_\infty = F(x_\infty) / qx_\infty$. Trajectories sufficiently close to the equilibrium will spiral in toward it. But the dynamics of motion distant from the equilibrium would seem to be entirely *ad hoc*, reflecting the arbitrary nature of the behavioral entry rule in [1.2].

Nevertheless, considerable attention has been given to theoretical study of the dynamics of this system, to determine whether trajectories sufficiently far from equilibrium might in principle lead to stock extinction. With the special linear structure of [1.2], the conclusion will depend very much on the prey growth function F and the effect of stock size on the effort-harvest relation.

The Smith model has been shown to exhibit considerable power in empirical analysis. Of particular interest to us here is a study by Wilen (1976) in which he re-examined the historical data from the late 19th century pelagic harvest of the Northern Pacific fur seal. At issue is whether that uncontrolled common property harvest would, if continued unchecked, have driven the fur seal stock to extinction. Wilen's study seems to suggest it would not have, but one is entitled to wonder to what extent these indications are model specific.

It is natural to wish to refine Smith's predator-prey model to incorporate non-equilibrium dynamics which, as in Gordon's original analysis of equilibrium, would be based more explicitly on the individual harvester's presumed profit-maximizing behavior. Such a refinement might, for one thing, provide a better insight into a bioeconomic system's "self-regulating" mechanisms. The key requirement is a proper characterization of entry into, and exit from, the active "predator" population.

Now, models of the **optimal** management of a harvested biological resource are well-established in the resource economics literature, and incorporate explicit profit-maximizing behavior on the part of the monopolistic sole owner or social manager. (See, for example, Clark 1976). The appropriate generalization, to analyze competitive exploitation by a number of independent firms, particularly of a common property resource, is more difficult in principle, and has been studied much less. (But see Levhari and Mirman 1980).

Of particular interest to us here is the model, due to Clark et al. (1978), of the sole owner exploitation of a marine population (fish or mammal), taking explicit account of "capital immalleability"; specifically, the irreversibility of major investment decisions. Capital immalleability is important in a modern, capital-intensive resource harvesting industry since it imparts a certain inertia to the bioeconomic system, through the so-called sunk-capital effect. The immalleable capital model has been applied by Clark and Lamberson (1982) in a normative study of the mid-century Antarctic whaling industry, presuming optimal management.

In this article, I develop a dynamic model of competitive exploitation of a common property renewable resource, assuming profit-maximizing goals by the individual participants in the harvest, and formulating their interaction as a differential game. Like Clark et al. (1978), I assume that irreversible investment decisions are important. With such an assumption, a natural rule for capital entry can be derived. This model is a generalization to an oligopolistic resource industry of an earlier model of mine, which addressed perfect competition (McKelvey 1985).

After deriving the model and noting some of its properties, I turn to applications. I re-examine the Clark and Lamberson (1982) analysis of Antarctic whaling, taking note that in its later phases the industry became in effect an oligopoly of five nations which negotiated harvest shares through the medium of the International Whaling Commission (I.W.C.).

Finally, I return to Wilen's (1976) analysis of the bioeconomics of fur seal harvest, where investment seems to have been almost completely reversible, but where other, shorter-term inertial factors seem to have played a role.

§2. The Model

I shall consider a fishing or whaling industry consisting of N firms, each operating a vessel or fleet of vessels, and each exerting harvest effort $\epsilon_n(t)$ within its effort capacity: $0 \leq \epsilon_n \leq \kappa_n$, for $n = 1, 2, ..., N$. Investment ι_n in capital capacity is irreversible, $0 \leq \iota_n$; i.e., disinvestment is ruled out. Taking account of depreciation,

[2.1] $\qquad d\kappa_n / dt = \iota_n - \gamma\kappa_n \ , \ \ 0 \leq \iota_n \ .$

Total industry rate of investment, total capacity, and total rate of effort are, respectively

$$I = \sum_{n=1}^{N} \iota_n \ , \qquad K = \sum_{n=1}^{N} \kappa_n \ , \qquad E = \sum_{n=1}^{N} \epsilon_n \ .$$

As is customary in modeling the harvesting of a wild marine stock, I shall assume that the individual firm's harvest rate $h_n(t)$ is proportional to its exerted effort, and also is proportional to the population size $x(t)$ of the harvested stock:

$$h_n = qx\epsilon_n \ .$$

Thus, the constraint on effort translates into a constraint on harvest:

$$0 \leq h_n \leq qx_n \ .$$

Total industry harvest rate is $H(t) = \displaystyle\sum_{n=1}^{N} h_n(t) \ .$

The objective of each firm is assumed to be to maximize the discounted stream of profits:

[2.2] $\pi_n = \displaystyle\int_0^\infty e^{-\delta t} \left[p(H) h_n - w_n \epsilon_n - c_n(I) \iota_n \right] dt$

$$= \int_0^\infty e^{-\delta t} \left\{ \left[p(H) - W_n(x) \right] h_n - c_n(I) \iota_n \right\} dt \ .$$

Here $W_n(x) = w_n / q_n x$ is the harvest cost rate per unit of catch, dependent on the current stock size x. Both price $p(H)$ and investment cost $c_n(I)$ may depend on total demand.

It remains to specify stock growth rate and how the biological stock responds to harvest. The simplest density dependent model (surplus production model) specifies that

[2.3] $\quad dx / dt = F(x) - H \ , \qquad 0 \leq H = qxE \leq qxK \ .$

Here $F(x)$ is the intrinsic growth rate for the population, assumed to depend only on current stock size. This formulation is broad enough to allow for a variety of differing characteristic responses of the population to the harvest. We shall examine cases in which $F(x)$ is, respectively, compensatory, depensatory, or over-depensatory (e.g., Clark 1976): These alternatives are illustrated in Figure 1. Prototypes for the three cases are, respectively,

$$F(x) = r\,x\,(1 - x/\check{x})\ , \text{ or}$$

$$= r\,x^{2+\alpha}\,(1 - x/\check{x})\ ,\ \alpha \geq 0\ , \text{ or}$$

$$= r\,x\,(1 - x/\check{x})\,(x/\underline{x} - 1)\ .$$

The response to a **constant** harvest effort rate E by each of these model variants is well-known (e.g., Clark 1976). For the compensatory model, every stock level $x^\# < \bar{x}$ can be achieved as a steady state, by maintaining effort $E = F(x^\#)/qx^\#$. This is a globally stable equilibrium. Only if an excessive effort is maintained ($E > sup\ F(x)/qx$) will the population be driven to extinction. In the depensatory case, there is a second unstable equilibrium $x_\#(E) < x^\#(E)$. Populations initially above $x_\#$ approach $x^\#$ under constant E, but populations below $x_\#$ will be driven to extinction. Finally, with over-depensation, extinction results for a population initially below $\underline{x}$ even if harvest effort ceases altogether.

A question before us is whether economic rationality will lead, for each of the three kinds of populations, to a timely slackening of harvest effort, leading to a sustained harvest rather than driving the population below critical levels and eventually to extinction.

Economic theory suggests that individual rationality ought to bring about a Nash (competitive) equilibrium among the firms in the bioeconomic system [2.1]-[2.3] (Owen 1982). This means that firms' individual harvest strategies will mesh, so that each firm achieves the maximum net return π_n **compatible with the actions of all the others**. Thus, no firm can unilaterally alter its harvest pattern without suffering losses.

Let us focus, then, on the typical individual firm n, and its harvest and investment strategy in response to **specified** harvest and investment patterns $h_m(t)$, $\iota_m(t)$ for all the other firms $m \neq n$. We shall use the notation

$$H = h_n + H_n, \text{ with } H_n = \sum_{m \neq n} h_m\ ;$$

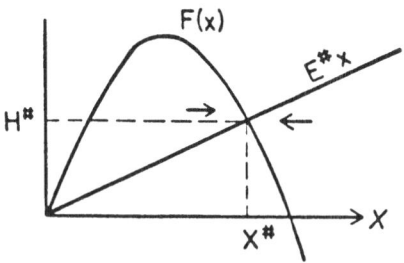

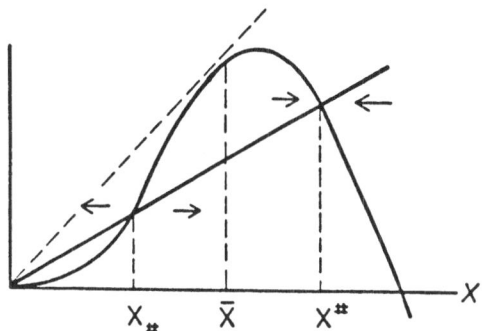

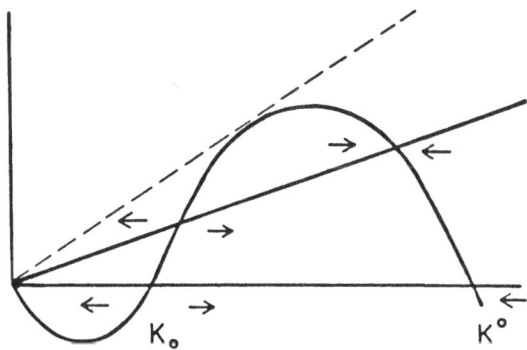

Figure 1. Growth curves $F(x)$: (a) compensatory, (b) depensatory, and (c) over-depensatory. Equilibrium under constant-effort harvest E is at intersection of the graph of $F(x)$ with the straight line qxE . Arrows show non-equilibrium direction of motion.

$$I = \iota_n + I_n, \qquad \text{with } I_n = \sum_{m \neq n} \iota_m .$$

Stated formally, the problem for firm n is: **Given** $H_n(t)$ and $I_n(t)$, to **choose** $h_n(t)$ and $\iota_n(t)$ so as to maximize π_n, under the constraints [2.1] and [2.3].

The problem may be reduced, using Mangasarian's (1966) version of the maximum principle. Here we shall simplify the analysis by limiting attention to an industry with firms that are alike in their unit harvest costs $W(x)$, and which face a constant price p and a constant and identical unit investment cost c. However, we return to the more general formulation in §5.

The current value Hamiltonian for firm n is

$$H_n = [\, p - W(x) \,] \, h_n - c \iota_n + \mu_n [\, \iota_n - \gamma \kappa_n \,] + \lambda_n [\, F(x) - h_n - H_n \,]$$

and the Lagrangian is

$$L_n = H_n + \sigma_n [\, q x \kappa_n - h_n \,] + \hat{\sigma}_n h_n + \rho_n \iota_n .$$

Here μ_n and λ_n are shadow prices for capital and biological stock, and the non-negative multipliers σ_n, $\hat{\sigma}_n$, and ρ_n are zero unless the corresponding constraint binds.

Differentiating L_n by κ_n and x yields the dual dynamic equations (in **current** value formulation):

[2.4] $d\mu_n / dt - \delta \mu_n = - \partial L_n / \partial \kappa_n = \gamma \mu_n - \sigma_n q x$,

and

[2.5] $d\lambda_n / dt - \delta \lambda_n = -\partial L_n / \partial x = - F'(x) \lambda_n + W'(x) h_n - \sigma_n q \kappa_n$.

The control variables h_n, ι_n are chosen so as to maximize H_n. In particular, setting $h_n^{max} = qx\kappa_n$ we find that

[2.6]
$$h_n = \begin{cases} 0 & \text{if} \quad p < W(x) + \lambda_n \\ h_n^{max} & \text{if} \quad p > W(x) + \lambda_n \end{cases} \quad .$$

Furthermore, when $p > W(x) + \lambda_n$ then the capacity constraint binds and $\sigma_n = p - W(x) - \lambda_n$; otherwise $\sigma_n = 0$. Hence, in every case

[2.7]
$$\sigma_n = [\, p - W(x) - \lambda_n \,]^+ \, ,$$

with $[\, y \,]^+ = max\,[\,0,y\,]$, indicating positive part. Similarly, by maximizing H_n over ι_n one finds that

[2.8]
$$\iota_n = \begin{cases} 0 & \text{if} \quad \mu_n < c \\ \infty & \text{if} \quad \mu_n > c \end{cases} \quad .$$

The latter alternative indicates an instantaneous pulse investment. An equilibrium state, where state variables remain constant over time, requires a constant finite investment rate and constant harvesting at capacity. Hence, from [2.8] and [2.6],

[2.9] (a) $\mu_n \equiv c$, (b) $W(x) + \lambda_n + \sigma_n \equiv p$.

Differentiating totally, using $dx\,/\,dt = 0$, yields

$$d\mu\,/\,dt \equiv 0, \qquad d\lambda_n\,/\,dt + d\sigma_n\,/\,dt \equiv 0 .$$

Substituting from equations [2.4] and [2.5] yield

[2.10]   (a) $\sigma_n = (\delta + \gamma)\, c / qx$,   (b) $\lambda_n = \dfrac{-W'_T(x) h_n}{\delta - F'(x)}$

where

$$h_n = q x \kappa_n = F(x) - H_n$$

and

$$W_T(x) = W(x) + (\delta + \gamma)\, c / qx = [w_o + (\delta + \gamma)\, c\,] / qx$$

is the total cost (including amortized investment cost) for operating at full capacity. Eliminating λ_n between [2.9b] and [2.10b] yields a characterization of the equilibrium x:

$$\lambda_n = h_n \; \frac{[\, p - W_T(x)\,]'}{\delta - F'(x)} = p - W_T(x) \; .$$

We now invoke symmetry: all firms react identically, hence have the **same** harvests and investments. Hence, $H = N h_n$, $I = N \iota_n$, $K = N \kappa_n$. Dropping subscripts on λ_n, μ_n, and σ_n, the dynamic equations become:

[2.11a] $dx / dt = F(x) - H$

[2.11b] $dK / dt = I - \gamma K$

[2.11c] $d\lambda / dt = (\delta - F')\, \lambda - \sigma q K / N + W'(x)\, H / N$

[2.11d] $d\mu / dt = (\delta + \gamma)\, \mu - \sigma q x$

where

[2.11e] $\sigma = [\,p - \lambda - W(x)\,]^+$

and

[2.11f] $H = \begin{cases} 0 & \text{when} \quad p < W(x) + \lambda \\ qxK & \text{when} \quad p > W(x) + \lambda \end{cases},$

[2.11g] $I = \begin{cases} 0 & \text{when} \quad \mu < c \\ \infty & \text{when} \quad \mu > c \end{cases}.$

The equilibrium stock level x^* and shadow price λ^* are characterized by

[2.12a] $\lambda^* = \dfrac{1}{N} F(x^*) \dfrac{[\,p - W_T(x^*)\,]'}{\delta - F'(x^*)} = p - W_T(x^*)$

while

[2.12b] $H^* = F(x^*), \quad I^* = K^* / \gamma = F(x^*) / qx^*, \quad \text{and} \quad \mu^* = c.$

Note that the case $N = 1$ corresponds to monopoly. The limit as $N \to \infty$ corresponds to rent-dissipating bioeconomic equilibrium: One simply sets $1/N = 0$ and $\lambda(t) \equiv 0$ in the preceding formulas. A direct development of the bioeconomic open-access fishery is to be found in McKelvey (1985).

Note that, in the bioeconomic limit, even though firms put a zero shadow value λ on the resource, they **do** anticipate future positive returns to harvest effort $[\,p > W(x)\,]$. However these returns serve only to balance off apportioned costs of prior capital investment $[\,p = W(x) + \sigma(t)\,]$. New capital investment at any time $t \geq 0$ is in response to a net positive real return, $\mu - c \geq 0$, for that asset, and may be thought of as entry by new firms (recall that $N = \infty$). These firms, according to [2.11d], must take into account intertemporal tradeoffs, as reflected in the discount rate δ: they do **not** simply put $\delta = \infty$.

§3. Phase Plane Analysis

The polar cases of $N = 1$ and $N = \infty$ have been analyzed by, respectively, Clark et al. (1978) and McKelvey (1984). The behavior of the model's solution trajectories for $1 < N < \infty$ will be intermediate between these.

Phase plane trajectories for $N = 1$ and $N = \infty$ are sketched in Figure 2, which assumes a compensatory growth function $F(x)$. In both cases, the trajectories spiral around the long-run equilibrium at (x^*, K^*). For a nascent industry there will be an initial pulse of investment to a relatively high level of capitalization. The larger the initial biological stock, the greater this capitalization level will be.

Harvesting then draws down the biological stock to well below the long-run equilibrium level. In fact, the stock level tends toward a middle-run "quasi-equilibrium" (x^0, K^0) which corresponds to free capital $(c = 0)$ in [2.12]. This is the "sunk-cost effect": once investment has been committed irreversibly, vessels will freely utilize the available capacity, and will continue to operate so long as **variable** costs (including shadow costs) at the margin do not exceed price p. This effect also implies that stock will not be drawn down **below** x^0, since that would entail a loss relative to variable costs, closing down the harvest (see equation [2.11f]).

By making the model non-linear in factor costs, one could replace pulse investment and the off-on switching of harvest effort by less abrupt transitions. Note also that the spread between x^* and x^0 (and thus the width of the spirals) results from the relative size of $W(x) = w_0 / qx$ and $W_T(x) = [w_0 + (\delta + \gamma) c] / qx$; i.e., the proportion of costs that represent amortized investments.

A comparison of the diagrams in the polar cases $N = 1$ and $N = \infty$ reveals two primary differences. First, the short and long run equilibrium positions are shifted, with higher capital investment and lower biological stock levels when $N = \infty$. Both of these are aspects of the "over-capitalization" that characteristically is associated with common property exploitation of a renewable resource.

Figure 3 shows how the model's solution trajectories are affected by a depensatory or over-depensatory growth function $F(x)$. The key conclusion is that, with depensatory F, though stock recovery from low levels will be slowed down, the stock eventually will recover. Only with over-depensatory F can stocks be driven to extinction, and then only for those particular trajectories, arising from high initial stock level, that fall below the critical level $\underline{x}$. This behavior is related to the feature of the model that imposes a complete cut-off of harvest when marginal variable costs rise above unit price.

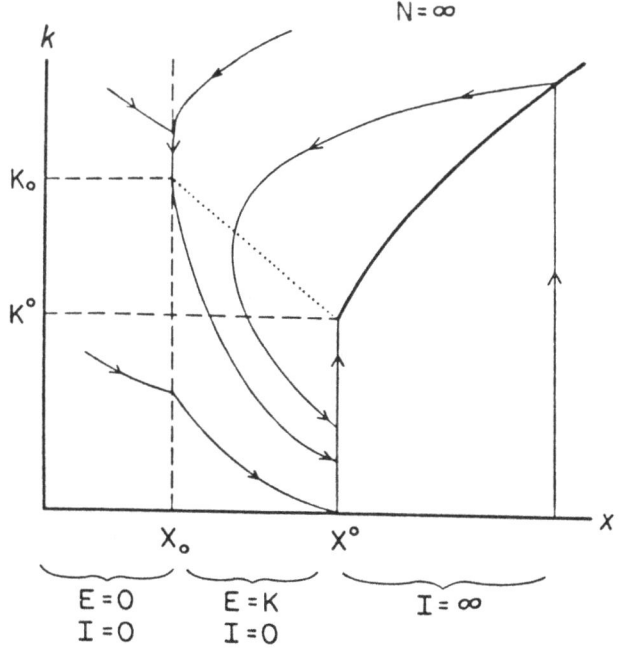

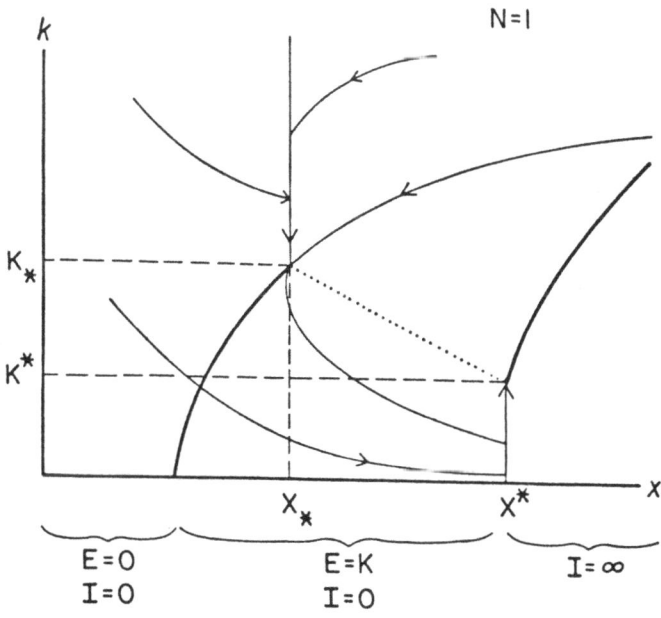

Figure 2. Phase-plane portrait: Compensatory model, (a) $N = \infty$ and (b) $N = 1$. (After, respectively, McKelvey 1985, and Clark et al. 1979).

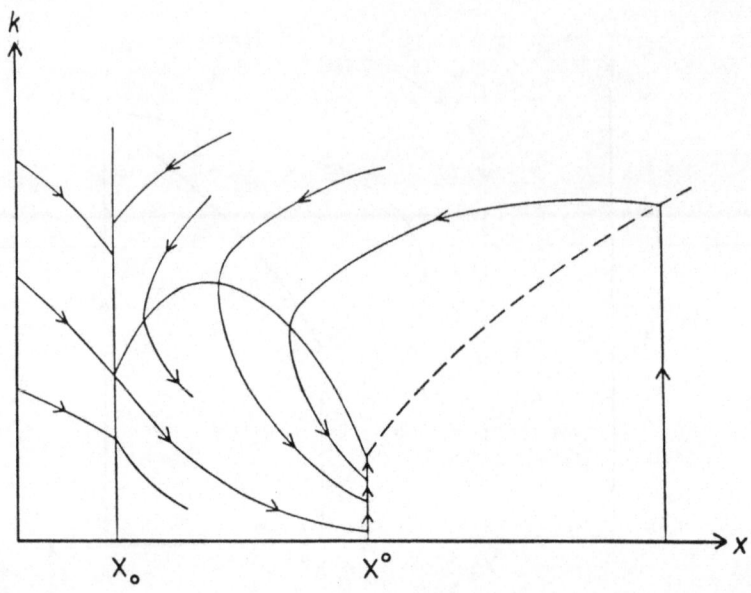

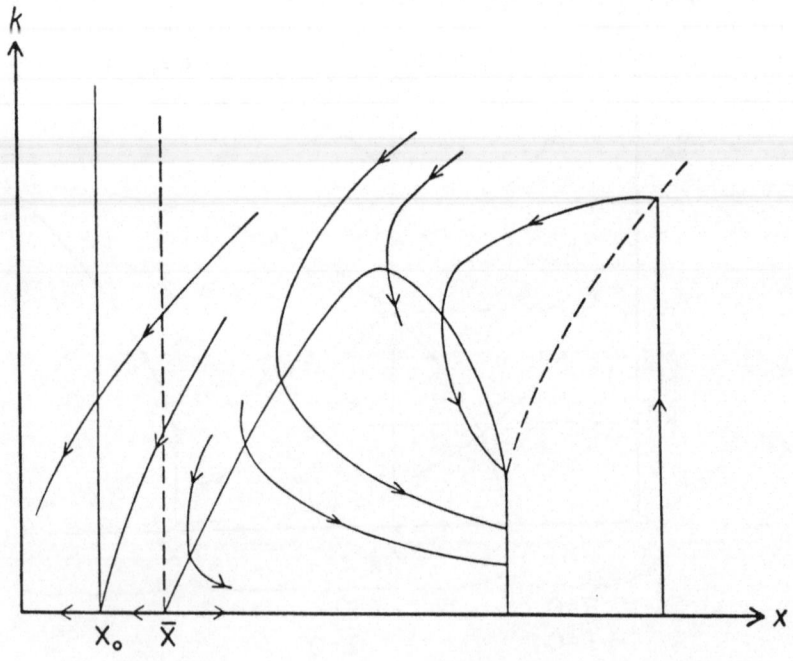

Figure 3. Phase-plane portrait: $N = \infty$; (a) depensation and (b) over-depensation.

Plainly, the conclusions of the last paragraph could be altered by straight-forward changes in the model, reflecting changed assumptions about nature and the economics of the harvest. Thus, for instance, the assumption that $H = qxE$; i.e., that harvest from a given effort drops linearly with the stock level, may be violated for certain species or environments, thereby making the stock more (or possibly less) vulnerable to harvest pressure, and thereby altering our conclusions about extinction (see Clark 1985).

§4. Application: The Great Whales

The effect of immalleability naturally is most significant in a capital intensive resource industry. This is the basis of the analysis of the mid-century Antarctic whaling industry by Clark and Lamberson (1982). Antarctic whale stocks, while heavily exploited beginning around 1925, were not harvested during World War II and had somewhat recovered by the end of the war. There then followed a 15-year period of rapid build-up of whaling fleets accompanied by an equally rapid decline of whale stocks. The fleet size peaked in 1961, with a total of 21 enormous factory vessels plus 261 catcher vessels, representing Japan, Norway, USSR, the United Kingdom, and the Netherlands. Thereafter, these countries succeeded in negotiating harvest quotas for each nation's fleet, through the International Whaling Commission (I.W.C.), and the following 20 years had been a period of relative stability of the size of the whale stock (though at a depressed level) and a steady decline in the size of the fleet.

Clark and Lamberson (1982) applied the Clark et al. (1979) model to examine what would have been the consequence of monopolistic control of the whaling industry. They assumed a logistic growth function

$$F(x) = r x (1 - x / \bar{x})$$

and made rough estimates of the biological and economic parameters. I have extended their analysis, assuming alternative industry structures: an open-access competitive whaling industry ($N = \infty$), and the Nash competitive equilibrium for common property exploitation by $N = 5$ nations. In our context, the monopolistic case ($N = 1$) may be presumed to forecast a cooperative solution achievable by the whaling nations, bargaining through the I.W.C.

The equilibrium stock level x^*_N for the multi-firm industry can be calculated by the following formula, which generalizes the standard formula for $N = 1$ with the logistic. In dimensionless form, let

$$Z_N^* = x_N^*/\bar{x} \qquad \text{and} \qquad \theta = \delta/\gamma$$

Then

$$Z_\infty = w_T/pq\bar{x} \qquad \text{and} \qquad Z_N = \frac{1}{4}\left[-B + \sqrt{B^2 + A}\,\right],$$

where

$$A = 8Z_\infty^*\left[\frac{1}{N} + \theta - 1\right], \qquad B = Z_\infty^*\left[\frac{1}{N} - 2\right] + \theta - 1.$$

Likewise, the "free capital" levels $Z^0{}_N = x^0{}_N/\tilde{x}$ are given by a parallel expression but with $w_T = w_0 + (\delta + \gamma)c$ replaced by w_0. Equilibrium capital levels are computed by $K_N^* = r(1 - Z_N^*)/q$.

Non-equilibrium characteristics are obtained by numerical integration of the differential system [2.11] working backwards from equilibrium at $t = \infty$. The results are shown graphically in phase-plane portraits, and key features, such as initial pulse investment K_{max}, and lowest whale stock level x_{min}, read off the curves.

The predictions of each of the models can be compared to the actual historical record of fleet size and evolution, and the corresponding (estimated) whale stock size. The results are presented in Figures 4, 5, and 6.

None of the models successfully simulates the initial 15-year capital build-up phase of the industry. This of course is an artifact of the model's linearity assumption concerning investment costs, which implies an instantaneous initial investment pulse. As I discuss in §5, one could do much better here by incorporating non-linear "adjustment costs". Indeed, the Smith (1969) phenomenological model simulates this build-up stage quite well, provided one "tunes" the *ad hoc* investment response factor β appropriately.

Possibly, the predicted size of the capital build-up is more significant than its phasing. In the cooperative model ($N = 1$) the predicted build-up is only about half of that actually observed. Both the common property ($N = 5$) and open-access ($N = \infty$) models double it.

It is during the subsequent period, of capital stock decline, that the models of immalleable investment ought to show their predictive superiority over the Smith (1969)

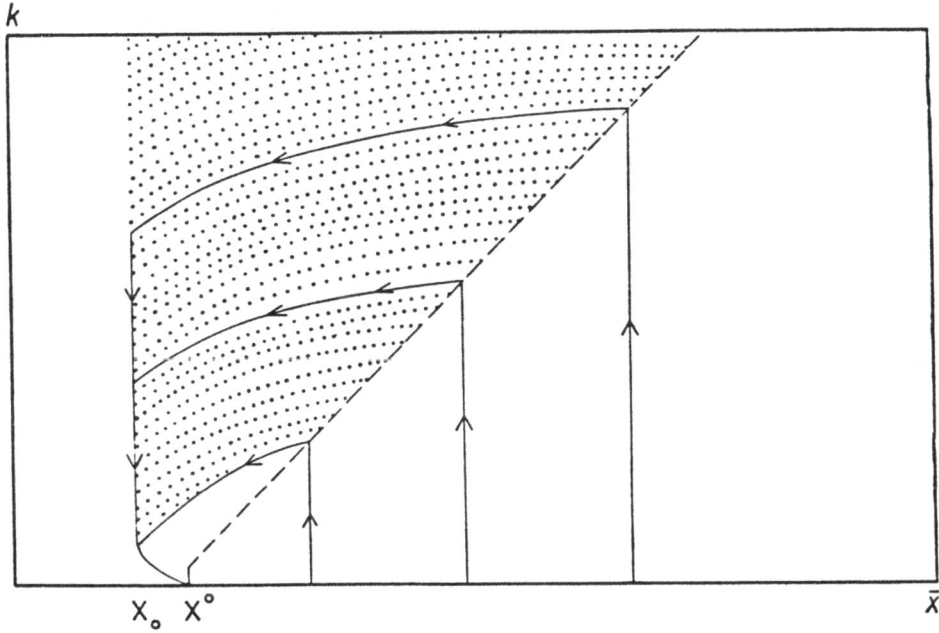

Figure 4. Open access whale harvest: $N = \infty$. Logistic model with parameter values from Clark and Lamberson (1982).

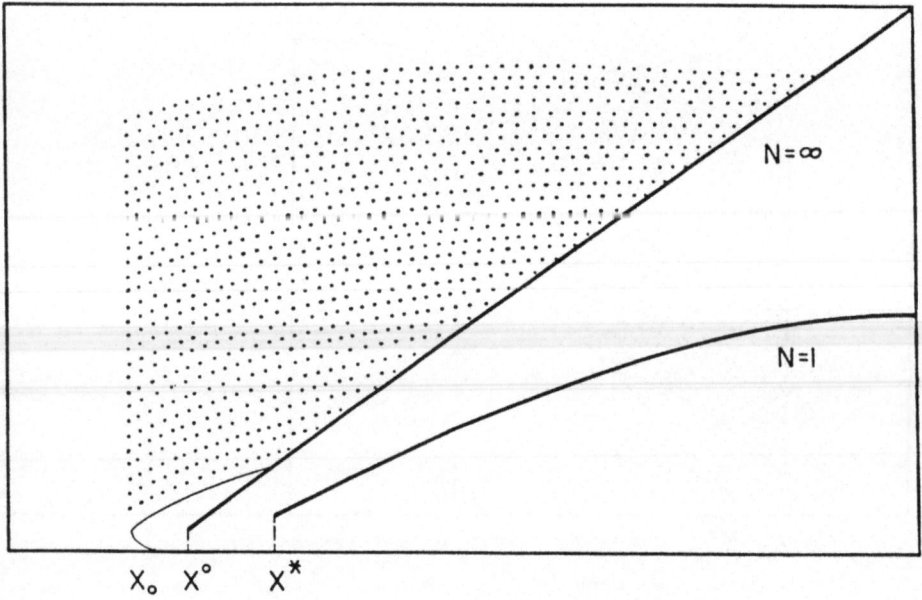

Figure 5. Threshold investment in a whale fleet as a function of initial whale stock level. $N = \infty$ vs $N = 1$. The logistic model is with parameter values from Clark and Lamberson (1982).

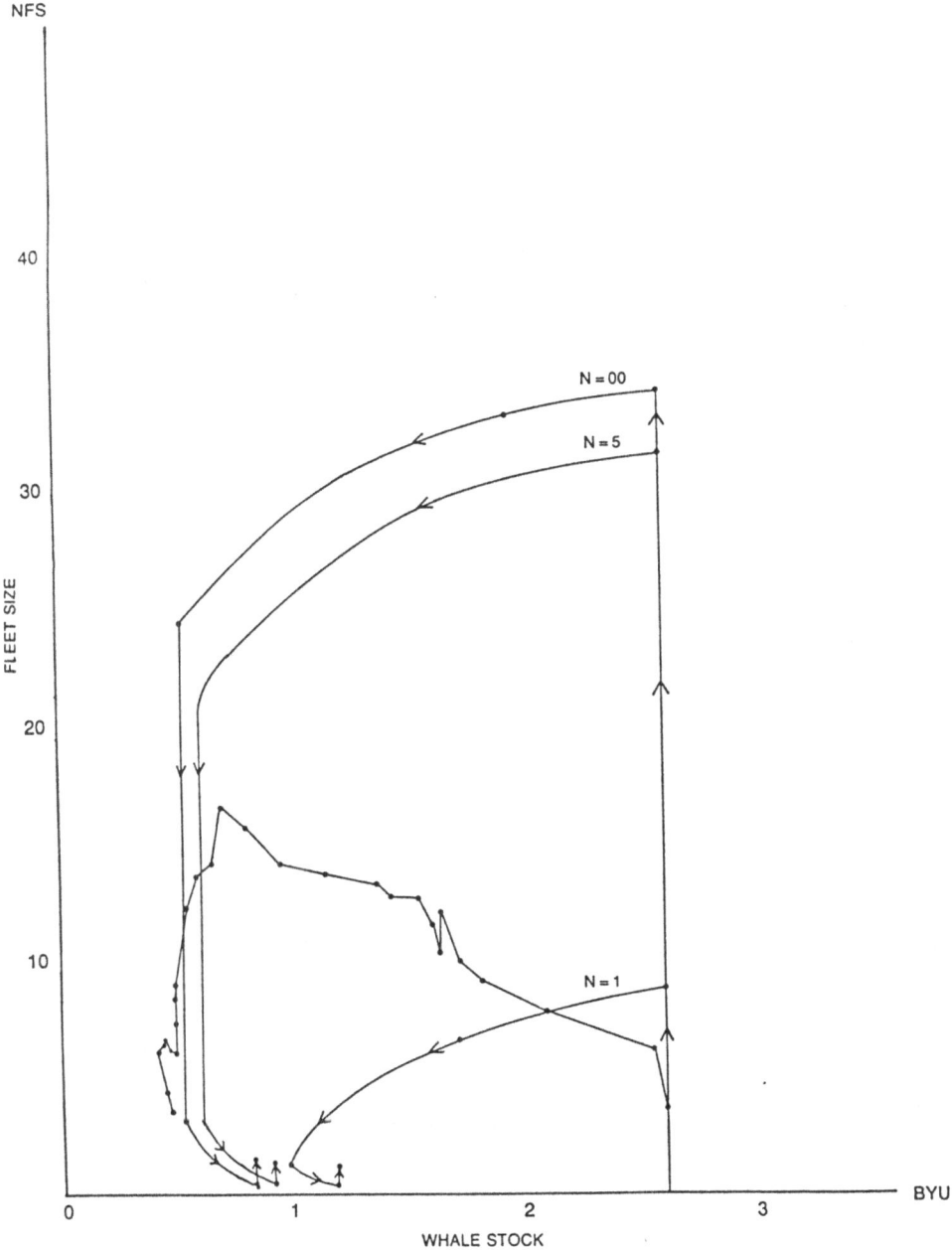

Figure 6. Rise and decline of the Antarctic Whaling Industry. Historical reconstruction is compared to model trajectories.

phenomenological model. (For example, they respond to short-term losses by immediately shutting down excess harvest capacity.) The fact that the whale stock level actually observed (or estimated) is nearly constant during this period suggests that we are near the steady state equilibrium. The predicted cooperative equilibrium stock level x_1^* seems much too high, compared with the observed stock level, but once again the common property and open access values are reasonably close. Of course, parameter values estimated in the models are very crude. As Clark has pointed out (1985), a lowered estimate of the whale stock's linear growth rate r would substantially lower predicted steady state stock levels.

The small difference between the results for $N = 5$ and $N = \infty$, as compared to $N = 1$, might be taken as indicating a very limited potential for achieving economic efficiency through limiting entry to 5 national fleets. But there was a substantial potential through I.W.C. negotiations. A comparison of the calculated cooperative solution ($N = 1$) with the realized historical trajectory suggests that this potential for cooperative advantage never was achieved.

Possibly the most striking feature of the model's behavior, and one which seems likely to be robust under model refinements, is the extent to which competitive common property practices serve to exaggerate the swings in the trajectory of resource utilization: not only is the final equilibrium state shifted (to the high capital, low resource-stock bionomic path), but the initial build-up of investment and the degree of overshoot to low resource-stocks both are **very** greatly exaggerated. This interaction, between sunk capital and common property exploitation, is likely to prove to be a pervasive feature of renewable resource utilization, wherever it occurs. (See for example McKelvey 1986). It is yet another pernicious aspect of the "tragedy of the common".

§5. Application: Pacific Fur Seal

In an imaginative study Wilen (1976), has investigated the economic history of the late 19th century open-access pelagic harvest of the Northern Pacific fur seal. A widespread fear at the time was that this intensive uncontrolled harvest on the open seas might have been driving the fur seal to extinction. Eventually economic losses due to the combined effects of falling prices and a precipitate decline in harvest success led to a period of more restrained, monopolistic harvesting, and ultimately the pelagic harvest was banned altogether.

Wilen (1976) re-examined the available historical data on prices, fleet size, and harvests, in the context of Smith's (1969) predator-prey model. He began by plotting the historical values of fleet size and (inferred) stock size on an x,K-phase plane, and then, by adjusting parameters in the dynamic equations [2.1] and [2.2] obtained a trajectory which gives a remarkably good fit. The adjustment involved tuning the unknown opportunity cost (which determines the equilibrium x_∞) and the reaction parameter β (which determines the tightness of the spiral).

Wilen's study suggests that bioeconomic factors might have combined so as to temper harvest enough to avoid extinction, though the historical experiment was interrupted by exogenous effects. So far as the model is concerned, one can prove for the Smith's model that over-depensation or strong depensation ($\alpha > 0$) are necessary in order for any trajectory to lead to extinction (see for example Berck 1979, Goh 1980).

In a rough way, the phenomenological model of Smith and the more mechanistic irreversible investment model developed in §2 are qualitatively similar in their solution dynamics. However, irreversible investment was **not** a significant factor in the Pacific fur seal harvest: according to Wilen, vessels could easily and rapidly disengage from sealing, by switching to the halibut fishery instead. And of course, our assumption of constant unit investment costs, and the resulting instantaneous capital build-up is totally unrealistic for Antarctic whaling, as well as for the pelagic seal harvest. (This point is further discussed in §4).

Aside from irreversible investment effects, there are at least two other specific economic forces that we ought to consider, which can introduce inertia to a model of a bioeconomic system. The first of these is imperfect competition: where the industry faces a downward sloping demand curve for its product. The second is non-linear investment costs, the so-called adjustment costs, of rapid investment or disinvestment. Adjustment costs provide a less extreme form of capital immalleability. Also, for the pelagic seal harvest, an appropriate idealization may be that of the open access competitive limit, $N = \infty$. Hence, the modeling framework developed in my earlier paper (McKelvey 1985) applies. Here, I shall only sketch the modifications needed in equations [2.11].

With market price $p = p(H)$, a decreasing function of total harvest H, the competitive harvest level is set by $\lambda \equiv 0$ and

$$p(H) = W(x) \quad ,$$

provided this equation has a solution in the range $0 \le H \le H_{max} = qxK$. Otherwise,

$$H = \begin{cases} 0 & \text{when} & p(0) < W(x) \\ H_{max} & \text{when} & p(H_{max}) > W(x) \end{cases}.$$

(This rule reduces to [2.11f] when applied to constant price p.) The function σ in [2.11e] becomes

$$\sigma = [\, p(H_{max}) - W(x) \,]^+ .$$

A caveat is required here. Wilen's (1979) data show that, in the pelagic seal harvest, prices and harvests rose and fell together around the peak period of the industry. However, it isn't clear that price was responding to harvest. In fact, on the way up, price led harvest, suggesting that the price rise was exogenous; i.e., $p = p(t)$.

To introduce adjustment costs we require that $C(I)$, the **total** cost of investment at rate I, be a monotone increasing and convex function: $C'(I) > 0$, $C''(I) > 0$ on $-\infty < I < \infty$, with $C'(0) = 0$. The slope $C'(I)$ might be discontinuous at $I - 0$, as one passes from disinvestment to investment.

The rate of investment is determined in the model to satisfy

$$\mu = C(I) / I = AC'(I) .$$

This is another manifestation of Gordon's (1954) bionomic principle: Open access investment rises to the level of break-even; there are no net profits.

The effect of adjustment costs will be to induce a gradual build-up of capital, and subsequent disinvestment as the biological stock level declines. As before, there will be some overshoot of the equilibrium. In a general way, this model's behavior parallels that in §3 and §4. The common property effects, on initial build-up and overstock, are the same.

Unfortunately, there seems to be no way to actually measure adjustment costs directly: all we can do is have recourse to curve-fitting, much in the way that Wilen has done. Still, the notion has a major conceptual advantage over Smith's (1969) phenomenological model in incorporating explicit profit-maximizing behavior of individual sealing vessels, and in allowing direct comparison of common property and monopolistic regimes.

In particular, the bioeconomic equilibrium level now is determined by

$$p(H) = W(x) + (\delta + \gamma) \, AC(I) / qx = W_T(x,I)$$

$$H = F(x) \ , \ I / \gamma = K = F(x) / qx \ ,$$

and this might serve to define an improved version of Smith's empirical "flexible accelerator" rule (see Nickell 1978).

There remains yet one other modeling device, that I wish to mention, as a means of introducing inertia to our bioeconomic system. This is the direct use of lags in the dynamic equations. Lags may be motivated by biological or physical constraints on the systems' rate of adjustment to changing circumstances, or may be regarded simply as providing a phenomenological description of system dynamics. So far as I know, no study has been made of the precise effect of lags in a model of immalleable investment in a common property resource industry.

It may be indeed that, aside from short term exogenous effects, what was occurring throughout this period was a straightforward driving-down of the stocks due to common property over-exploitation. In fact, the pelagic harvest did continue beyond the period modeled by Wilen. In the late 1890's a Japanese fleet entered the Bering Sea and Northwest Coast sealing grounds, and remained active until the 1911 Treaty stopped all pelagic sealing. A simple plot (Figure 7) of catch-per-effort (an index of stock abundance) against time through that period shows a seemingly inexorable decline. In this representation the stock recovery observed in the late 1980's looks very much like a short-term statistical aberration.

Which interpretation is right? Probably the evidence isn't adequate to decide. Still, from a theoretical perspective, a turn around in seal stock levels has to be tied to some inertial effects present, and these have not been identified. In their absence one must conclude that the "tragedy of the common" surely was being played out yet again.

Acknowledgments. I wish to thank Ronald Lamberson for providing me with historical information about the post-World War II whaling industry, and Colin Clark for insightful comments. I am indebted to William Derrick for the computer simulations

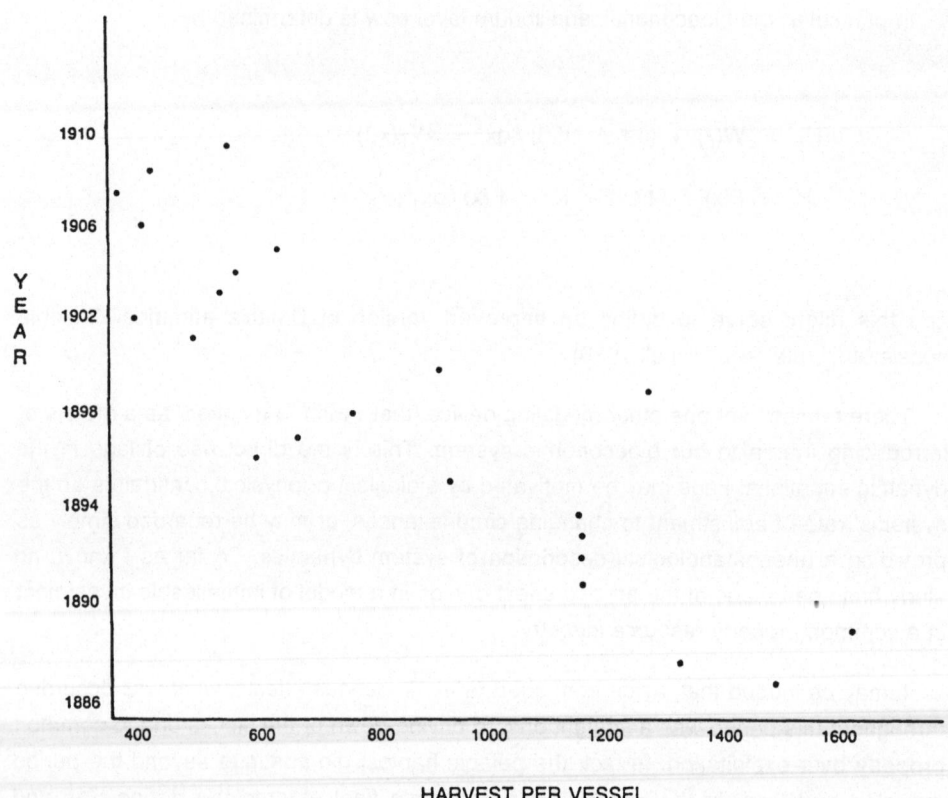

Figure 7. The Pelagic Seal Harvest. Harvest per vessel vs time: 1886-1909.
(Calculations based on data quoted by Wilen 1976).

shown in Figures 4 and 5. Financial support was provided by NSERC grant A-3990, during a period of residency at the University of British Columbia.

Literature Cited

Berck, P. 1979. Open Access and extinction. *Econometrica* 47: 877-882.

Clark, C.W. 1976. *Mathematical Bioeconomic: The Optimal Management of Renewable Resources.* Wiley, New York.

Clark, C.W. 1985. *Bioeconomic Modeling and Fisheries Management.* Wiley, New York.

Clark, C.W., F.H. Clarke, and G.R. Munro. 1979. The optimal exploitation of renewable resource stocks: Problems of irreversible investment. *Econometrica* 47: 25-49.

Clark, C.W., and R.H. Lamberson. 1982. An economic history and analysis of pelagic whaling. *Marine Policy* 6: 103-120.

Goh, B.S. 1980. *Management and Analysis of Biological Populations.* Elsevier, New York.

Gordon, S. 1954. The economic theory of a common property resource: The fishery. *Journal of Political Economics* 62: 124-142.

Levhari, D., and L.J. Mirman. 1980. The great fish war: An example using a dynamic Cournot-Nash solution. *Bell Journal of Economics* 11: 322-344.

Mangasarian, O.L. 1966. Sufficient conditions for the optimal control of nonlinear systems. *SIAM Journal of Control* 4: 139-152.

McKelvey, R. 1985. Decentralized regulation of a common property renewable resource industry with irreversible investment. *Journal of Environmental Economics and Management* 12: 287-307.

McKelvey, R. 1986. Groundwater-based agriculture in the arid American West: Modeling the transition to a steady-state renewable resource industry. *Lecture Notes in Biomathematics* . In press.

Nickell, S.J. 1978. *The Investment Decisions of Firms.* Cambridge University Press, Cambridge, England.

Owen, G. 1982. *Game Theory.* Academic Press, New York. 2nd Edition.

Smith, V.L. 1969. On models of commercial fishing. *Journal of Political Economics* 77: 181-198.

Wilen, J.E. 1976. Common property resources and the dynamics of over-exploitation: The case of the North Pacific fur seal. *University of British Columbia, Vancouver, Economics Department Resource Paper Number 3.*

PREDATOR-PREY COEVOLUTION AS AN EVOLUTIONARY GAME

Joel S. Brown

Department of Ecology and Evolutionary Biology

and

Thomas L. Vincent

Department of Aerospace and Mechanical Engineering
University of Arizona, Tucson, AZ 85721 USA

Abstract.   To model evolution as an evolutionary game, we have used a fitness generating function to define the fitness of any individual in a community bounded by the same evolutionary constraints. Using a single fitness generating function, we have previously investigated the effect of external inputs on a community at an evolutionarily stable strategy (ESS). Of particular interest are the circumstances under which the external input promotes the coexistence of several strategies in a community that otherwise would have a single-strategy ESS. The external inputs can include physiographic changes, human intervention, or the introduction of a new species not modeled by the single fitness generating function. We consider in detail here the situation of introducing a predator into a hitherto unexploited community of prey. In this case, the prey and predators each have a separate set of evolutionary constraints which produce two different fitness generating functions. Necessary conditions for determining the ESS under two or more fitness generating functions are presented. The coevolution of predator and prey is then examined with the aid of frequency-dependent adaptive landscapes, one for each fitness generating function. As a result of disruptive selection imposed by the predator, we obtain an ESS composed of two coexisting prey strategies and a single predator strategy.

§1. Introduction

Previously, in our development of a theory for evolution as a continuous game (Vincent and Brown 1984, Vincent 1985, Brown and Vincent 1986), we assumed that all individuals shared the same set of evolutionary feasible phenotypes and experienced the same ecological consequences of those phenotypes. We used a single fitness generating function to model the evolutionary process. However, this may not be realistic for whole communities, where individuals from distantly related taxa cannot be represented by the same fitness generating function or by the same set of feasible phenotypes.

Here, we extend our theory of evolutionary games to include coevolution among and between individuals of different taxa. Our development allows for any number of fitness generating functions. The evolutionarily stable strategy (ESS) may be a coalition of strategies. In fact, the ESS coalition may represent several strategies from each fitness generating function. Coexistence of strategies may result from within or between fitness generating functions. We illustrate the theory by modeling predator-prey coevolution. We assume that the strategy used by an individual predator or prey influences its competitive interactions with other predators or prey. In addition, the individual's strategy influences its ability to capture prey or avoid predation.

Others have developed theories for modeling predator-prey coevolution. Schaffer and Rosenzweig (1978) and Rosenzweig and Schaffer (1978) model coevolution as a "Red Queen." That is, both predator and prey are constantly evolving and acquiring new mutations to assist them in capturing prey or avoiding predators. While an ESS is never achieved under the Red Queen hypothesis, the interaction between predator and prey may achieve a steady state. Roughgarden (1983) uses a theory of evolution to model predator-prey coevolution. It defines the evolutionary steady state as the set of strategies that maximize population fitness. The approach is inappropriate for models involving frequency-dependent selection (see Brown and Vincent 1987). Our model of predator-prey coevolution differs from Schaffer and Rosenzweig's (1978) in that the predators and prey achieve an ESS; it differs from Roughgarden's (1983) in that it includes frequency-dependent selection.

Paine (1966) showed that the coexistence of several intertidal organisms was predator mediated; removal of the predator resulted in the competitive exclusion of several prey species. Our example documents a notably different means by which a predator can enhance the diversity of competitors. In our model, the predator, by attempting to maximize its fitness, will exert disruptive selection on a prey population. The resulting coevolutionary process results in the coexistence of two prey strategies. While the resultant prey strategies will coexist in the absence of the predator, the two strategies are not evolutionarily stable in the absence of the predator.

§2. A Theory for Species Coevolution

Consider a community of coevolving organisms. Each individual has a number of traits which exhibit heritable variation. The phenotype of an individual denotes the specific value for each trait. The individual does not choose its phenotype but, rather, inherits it from its parent(s). We will assume the simplest genetic transition mechanism. Each phenotype reproduces itself as an asexual clonal species. Under this assumption, a phenotype defines a species. New phenotypes arise as rare mutations. The fitness of an individual is not only influenced by the phenotypes of others, but, in general, the fitness maximizing phenotype of an individual is also influenced by others' phenotypes. In modeling coevolution as a game, the individuals are the players and their phenotypes are the strategies.

The strategy of the ith individual in the community is an n-dimensional vector

[2.1] $$u^i = \left[u^i_1, \dots, u^i_n \right] ,$$

where n is the total number of traits which completely identify the individual. An element of the vector denotes a specific trait, and the value of that element is the actual phenotype of the individual at that trait. The traits may be behavioral (aggressiveness), morphological (body size), or physiological (digestive efficiency).

Let $r(t)$ be the total number of distinct strategies found among individuals in the community at time t, and let u be the row vector of all such strategies

[2.2] $$\tilde{u} = \left[u^1, \dots, u^r \right] .$$

Note that we denote the component of a strategy by a subscript and a particular strategy vector by a superscript. In general, each strategy vector u^i, $i = 1, \ldots, r$, lies in some subset of the n-dimensional characteristic space; that is

[2.3] $u^i \in U_i \subseteq E^n$,

where U_i is the strategy set of individual i. The set U_i represents the particular subset of n-dimensional Euclidean space (E^n) that restricts u^i. The strategy set of an individual can be thought of as the set of feasible mutations which can (and will) arise from the individual's lineage.

Let N_i be the population density of individuals using strategy u^i, and let N be the row vector of population densities

[2.4] $N = [N_1, \ldots, N_r]$.

Let $H_i(\tilde{u}, N)$ be the fitness of individuals using strategy u^i, and let the dynamics of this community be given by

[2.5] $N_i(t + 1) = N_i(t) H_i(\tilde{u}, N)$ $i = 1, \ldots, r$

where $N_i(t + 1)$ refers to the population density of N_i at time $t + 1$. [Note that in this paper we restrict ourselves to difference equations and discrete time; see Vincent and Fisher (1987) for an extension of the theory to continuous time.]

For what follows, we need to identify certain strategies in the population as being separate from the rest. Let the vector

[2.6] $u^o = [u^1, \ldots, u^\sigma]$

denote the first $\sigma \leq r$ strategies of $\tilde{u}$, and let

$$[2.7] \qquad u^m = [u^{\sigma+1}, \dots, u^r]$$

denote the remaining (mutant) strategies. It follows that

$$[2.8] \qquad \tilde{u} = [u^o, u^m] \ .$$

Let us now define the following ordering for the population vector N:

$$[2.9] \qquad
\begin{array}{llll}
N = 0 & \text{if and only if } N_i = 0 & i = 1, \dots, r \\[2mm]
N > 0 & \text{if and only if } N_i > 0 & i = 1, \dots, r \\[2mm]
N \succsim 0 & \text{if and only if } N_i > 0 & i = 1, \dots, \sigma \\[2mm]
 & \qquad\qquad\quad\ N_i = 0 & i = \sigma+1, \dots, r
\end{array}$$

Definition: A vector u^o composed of the strategy vectors $u^i \in U_i$, $i = 1, \dots, \sigma$, is said to be a <u>coalition vector</u> if there exists a population vector $N^* \geq 0$ such that for all $u^i \in U_i$, $i = \sigma+1, \dots, r$, and for all $N(0) \succsim 0$, equation [2.5] yields

$$[2.10] \qquad \lim_{t \to \infty} N(t) = N^* \ .$$

It follows from the definition that, at the equilibrium point N^*, all strategies in the coalition vector must have a fitness of one:

$$[2.11] \qquad H_i(\tilde{u}, N^*) = 1 \ .$$

The fraction of the total population using strategies in the coalition vector is called the coalition frequency, P_0, where

[2.12] $P_o = \sum_{i=1}^{\sigma} \dfrac{N_i(t)}{N(t)}$.

Definition: A coalition vector u^o is said to be an ESS if there exists a generation time $t_m \geq 0$ such that for all strategy vectors $u^i \in U_i$, $i = \sigma+1, \ldots, r$, and all $N(0) > 0$, equations [2.5] and [2.12] yield a monotone increasing sequence for the coalition frequency P_0 when $t = t_m, t_{m+1}, \ldots$, and

[2.13] $\lim_{t \to \infty} N(t) = N^* \to \lim_{t \to \infty} P_o(t) = 1$.

We have previously shown (Brown and Vincent, 1986, lemma 2.1) that if u^o is an ESS with $r > \sigma$, then as $t \to \infty$, the fitness associated with any strategy in the coalition u^o must be greater than the average fitness of the remaining (mutant) strategies. That is,

[2.14] $H_j(\tilde{u}, N) > \bar{H}_m$ for $j = 1, \ldots, \sigma$

where

[2.15] $\bar{H}_m = \dfrac{\sum_{i=\sigma+1}^{r} N_i(t) H_i(\tilde{u}, N)}{\sum_{i=\sigma+1}^{r} N_i(t)}$

is the average fitness of the mutants. Since this condition must hold for any number of mutant strategies, it must hold in particular for one mutant. In this case, [2.14] simplifies and we obtain

[2.16] $\qquad H_j(\tilde{u}, N) > H_{\sigma+1}(\tilde{u}, N) \qquad$ for $j = 1, \dots, \sigma$.

Since in the limit $N(t) \underset{t \to \infty}{\longrightarrow} N^*$ and $H_j(\tilde{u}, N^*) = 1$, we have

[2.17] $\qquad H_{\sigma+1}(\tilde{u}, N^*) \leq H_j(\tilde{u}, N^*) = 1 \qquad$ for $j = 1, \dots, \sigma$.

This result is made more useful with the definition of evolutionary identical individuals.

Definition: Individual organisms identified by the strategies u^i and u^j are said to be evolutionarily identical if they draw their strategies from the same strategy set, i.e., $U_i \equiv U_j$ and

$$ H_i(\tilde{u}, N) = H_j(\tilde{u}, N) $$

for all $u_i = u_j$.

Two individuals are evolutionarily identical if they share the same strategy set and if the ecological consequences of those strategies are the same for both individuals. The strategy sets of two individuals must be equivalent in both character and mathematical dimension. That is, individuals that are evolutionarily identical must possess the same traits and the same range of evolutionary feasible values for these traits. For example, body length and wing length may be represented by a scalar over (perhaps) the same range. However, since they are different traits, they differ in character.

In what follows, we assume that there are s groups of evolutionarily identical individuals and therefore s distinct strategy sets, U_i, where $i = 1, \dots, s \leq \sigma$. Note that we have assumed that the maximum number of groups is less than or equal to the number of strategies in the coalition vector. Furthermore, among the individuals of the ith evolutionarily identical group, there are r_i strategies present. Without any loss of generality, we use the following procedure to assign strategies according to group. Let

[2.18] $u(i,j) = u^k$,

where

[2.19] $k = j + \displaystyle\sum_{m=0}^{i-1} r_m \qquad \begin{array}{l} i = 1, \dots, s \\ j = 1, \dots, r_i \text{ and } r_o = 0 \end{array}$.

According to [2.18] and [2.19], the total strategy vector $\tilde{u}$ is partitioned according to evolutionarily identical groups. The first r_1 strategies correspond to those present within the first group, the second r_2 strategies correspond to group 2, etc. Thus, $u(i,j)$ is the jth strategy of group i. All individuals of group i must draw from the appropriate strategy set

[2.20] $u(i,j) \in U_i \qquad \begin{array}{l} i = 1, \dots, s \\ j = 1, \dots, i_i \end{array}$.

In a fashion similar to our previous work, we may now define a fitness generating function for each evolutionary identical group.

Definition: A function $G_i(u, \tilde{u}, N)$ is said to be a fitness generating function for an evolutionary identical group i if

[2.21] $G_i[u(i,j), \tilde{u}, N] \equiv H_k(\tilde{u}, N)$,

where k is defined in equation [2.19]. According to [2.21], G_i is a fitness generating function for group i if the fitness of individuals of group i using strategy $u(i,j)$ is obtained by replacing u in G with $u(i,j)$.

Let $0 \leq \sigma_i \leq \sigma$, $i = 1, \dots, s$, be the number of strategies associated with each evolutionarily identical group in an ESS vector. It now follows that condition [2.17] may be written in terms of fitness generating functions as follows:

[2.22] $G_i [u(i,\sigma+1) , \tilde{u} , N^*] \leq G_i [u(i,j) , \tilde{u} , N^*] = 1$

for $i = 1 , ... , s$; $j = 1 , ... , \sigma_i$. This result is summarized in the following theorem.

Theorem: Let u^0 be a coalition vector with the equilibrium density vector $N^* \gtrsim 0$. Let $G_i(u , \tilde{u} , N)$, $i = 1 , ... , s$, be the fitness generating function for each evolutionary identical group i in the population. If u^0 is an ESS under the dynamics given by [2.5], then for each $i = 1 , ... , s$, the function

[2.23] $G_i(u , \tilde{u} , N^*)$

must take on a global maximum with respect to $u \in U_i$ at $u(i,j)$ for $j = 1 , ... , \sigma_i$. The maximum value of G_i at each of these points equals 1.

In the process of actually formulating the G function, it will usually be necessary to identify the number of individuals using strategy j of group i. We will use the same notation as we did to assign strategies. That is, $N(i,j)$ will refer to the density of individuals using strategy j of group i.

§3.   Predator-prey Coevolution

Consider now a predator-prey interaction. We assume that the prey belongs to one evolutionarily identical group and the predators to another. Let $u(1,i) \in U_1$ be the ith strategy of the prey and $u(2,i) \in U_2$ be the ith strategy of the predator. The corresponding densities for each of these strategies are designated by $N(1,i)$ and $N(2,i)$. Assume that the number of prey individuals in the population using strategy $u(1,i)$ in the next generation is given by

$$[3.1] \quad N_{t+1}(1,i) = N_t(1,i) \left\{ 1 + R_1 \left\{ 1 - \frac{\displaystyle\sum_{j=1}^{r_1} \alpha[u(1,i), u(1,j)] N(1,j)}{K[u(1,i)]} \right\} \right.$$

$$\left. - \sum_{j=1}^{r_2} \beta[u(1,i), u(2,j)] N(2,j) \right\} ,$$

and the number of predator individuals in the population using strategy $u(2,i)$ in the next generation is given by

$$[3.2] \quad N_{t+1}(2,i) = N_t(2,i) \left\{ 1 + R_2 \left\{ 1 - \frac{\displaystyle\sum_{j=1}^{r_2} N(2,j)}{b \displaystyle\sum_{j=1}^{r_1} \beta[u(1,j), u(2,i)] N(1,j)} \right\} \right\} .$$

Recall that r_1 and r_2 are the total number of strategies associated with the prey and predators, respectively. The intrinsic growth rates are designated by R_1 and R_2. The α function represents the competitive interactions among the prey. The function β is the effectiveness of the predators at harvesting the prey. The K function is associated with the carrying capacity of the prey in the absence of predators and competitors. Finally, b is the efficiency factor associated with the conversion of prey into predators.

Using [3.1], we obtain the following fitness generating function for the prey:

$$[3.3] \quad G_1(u, \tilde{u}, N) = 1 + R_1 \left\{ 1 - \frac{\displaystyle\sum_{j=1}^{r_1} \alpha[u, u(1,j)] N(1,j)}{K(u)} \right\} - \sum_{j=1}^{r_2} \beta[u, u(2,j)] N(2,j)$$

Using[3.2], we obtain the following fitness generating function for the predators:

$$[3.4] \qquad G_2(u, \tilde{u}, N) = 1 + R_2 \left\{ 1 - \frac{\displaystyle\sum_{j=1}^{r_2} N(2,j)}{\displaystyle\sum_{j=1}^{r_1} \beta[u(1,j), u] N(1,j)} \right\} .$$

In seeking an ESS, the necessary conditions as obtained from the theorem will depend upon the number of strategies in the coalition vector. Since we do not know a priori the number of strategies in the coalition vector, we begin by seeking an ESS with a coalition of one prey and one predator strategy and see if we obtain multiple solutions (or no solutions). If an ESS coalition of one prey and one predator strategy does not exist, we then seek a coalition of two prey and one predator, two prey and two predators, etc.

In this section, we first develop, for the above model, necessary conditions for a coalition of one prey and one predator strategy and then develop necessary conditions for a coalition of two prey and one predator strategy. Necessary conditions for other combinations involve similar procedures (see, for example, Vincent and Brown 1986, for the development of conditions of one and two "prey" strategies with no predators).

First assume a coalition of one prey and one predator strategy. The appropriate G function for use with the theorem are given by

$$[3.5] \qquad G_1(u, \tilde{u}, N^*) = 1 + R_1 \left\{ 1 - \frac{\alpha[u, u(1,1)] N^*(1,1)}{K(u)} \right\} - \beta[u, u(2,1)] N^*(2,1)$$

$$[3.6] \qquad G_2(u, \tilde{u}, N^*) = 1 + R_2 \left\{ 1 - \frac{N^*(2,1)}{b\,\beta[u(1,1), u] N^*(1,1)} \right\} .$$

Since the strategy sets are unbounded, necessary conditions to maximize G_1 at $u(1,1)$ and G_2 at $u(2,1)$ are given by

$$[3.7] \qquad \frac{\partial G_1[u(1,1), \tilde{u}, N^*]}{\partial u} = 0$$

and

[3.8] $$\frac{\partial G_2 \, [\, u(2,1) \, , \, \tilde{u} \, , \, N^* \,]}{\partial u} = 0 \, ,$$

with the equilibrium conditions

[3.9] $$G_1 \, [\, u(1,1) \, , \, \tilde{u} \, , \, N^* \,] = 1$$

and

[3.10] $$G_2 \, [\, u(2,1) \, , \, \tilde{u} \, , \, N^* \,] = 1 \, .$$

With the *G* functions given by [3.5] and [3.6], four conditions [3.7]-[3.10], in turn, yield

[3.11] $$- \frac{\partial \beta \, [\, u(1,1) \, , \, u(2,1) \,]}{\partial u} \, N^*(2,1) \, \frac{K \, [\, u(1,1) \,]^2}{R_1} =$$

$$K \, [\, u(1,1) \,] \, \frac{\partial \alpha \, [\, u(1,1) \, , \, u(1,1) \,]}{\partial u} - \frac{\partial K \, [\, u(1,1) \,]}{\partial u} \, \alpha \, [\, u(1,1) \, , \, u(1,1) \,] \, N^*(1,1)$$

[3.12] $$\frac{\partial \beta \, [\, u(1,1) \, , \, u(2,1) \,]}{\partial u} = 0$$

[3.13] $$R_1 \left\{ 1 - \frac{\alpha \, [\, u(1,1) \, , \, u(1,1) \,] \, N^*(1,1)}{K \, [\, u(1,1) \,]} \right\} = \beta \, [\, u(1,1) \, , \, u(2,1) \,] \, N^*(2,1)$$

[3.14] $$b \, \beta \, [\, u(1,1) \, , \, u(2,1) \,] \, N^*(1,1) = N^*(2,1) \, .$$

Consider now a coalition of two prey and one predator strategy. The appropriate G functions for use with the theorem are given by

$$[3.15] \qquad G_1(u, \tilde{u}, N^*) = 1 + R_1 \left\{ 1 - \frac{\alpha[u, u(1,1)]N^*(1,1) + \alpha[u, u(1,2)]N^*(1,2)}{K(u)} \right\}$$

$$- \beta[u, u(2,1)]N^*(2,1)$$

$$[3.16] \qquad G_2(u, \tilde{u}, N^*) = 1 + R_2 \left\{ 1 - \frac{N^*(2,1)}{b\beta[u(1,1), u]N^*(1,1) + b\beta[u(1,2), u]N^*(1,2)} \right\} .$$

Again, since the strategy sets are unbounded, the necessary conditions to maximize G_1 at $u(1,1)$ and $u(1,2)$ and G_2 at $u(2,1)$ are given by

$$[3.17] \qquad \frac{\partial G_1[u(1,1), \tilde{u}, N^*]}{\partial u} = 0$$

$$[3.18] \qquad \frac{\partial G_1[u(1,2), \tilde{u}, N^*]}{\partial u} = 0$$

$$[3.19] \qquad \frac{\partial G_2[u(2,1), \tilde{u}, N^*]}{\partial u} = 0 ,$$

with the equilibrium conditions

$$[3.20] \qquad G_1[u(1,1), \tilde{u}, N^*] = 1$$

$$[3.21] \qquad G_1[u(1,2), \tilde{u}, N^*] = 1$$

[3.22] $\quad G_2 [\, u(2,1) \,,\, \tilde{u} \,,\, N^* \,] = 1 \,.$

With the G functions given by [3.5] and [3.6], these latter six conditions, in turn, yield

[3.23] $\quad - N^*(2,1) \dfrac{K [\, u(1,1) \,]^2}{R_1} \dfrac{\partial \beta [\, u(1,1) \,,\, u(2,1) \,]}{\partial u} =$

$K [\, u(1,1) \,] \left\{ N^*(1,1) \dfrac{\partial \alpha \,[\, u(1,1) \,,\, u(1,1) \,]}{\partial u} + N^*(1,2) \dfrac{\partial \alpha \,[u(1,1) \,,\, u(1,2) \,]}{\partial u} \right\}$

$- \dfrac{\partial K [\, u(1,1) \,]}{\partial u} \left\{ \alpha \,[\, u(1,1) \,,\, u(1,1) \,]\, N^*(1,1) + \alpha \,[\, u(1,1) \,,\, u(1,2) \,]\, N^*(1,2) \right\}$

[3.24] $\quad - N^*(2,1) \dfrac{K [\, u(1,2) \,]^2}{R_1} \dfrac{\partial \beta [\, u(1,2) \,,\, u(2,1) \,]}{\partial u} =$

$K [\, u(1,2) \,] \left\{ N^*(1,1) \dfrac{\partial \alpha \,[\, u(1,2) \,,\, u(1,1) \,]}{\partial u} + N^*(1,2) \dfrac{\partial \alpha \,[\, u(1,2) \,,\, u(1,2) \,]}{\partial u} \right\}$

$- \dfrac{\partial K [\, u(1,2) \,]}{\partial u} \left\{ \alpha \,[\, u(1,2) \,,\, u(1,1) \,]\, N^*(1,1) + \alpha \,[\, u(1,2) \,,\, u(1,,2) \,]\, N^*(1,2) \right\}$

[3.25] $\quad N^*(1,1) \dfrac{\partial \beta [\, u(1,1) \,,\, u(2,1) \,]}{\partial u} + N^*(1,2) \dfrac{\partial \beta [\, u(1,2) \,,\, u(2,1) \,]}{\partial u} = 0$

[3.26] $\quad R_1 \left\{ 1 - \dfrac{\alpha \,[\, u(1,1) \,,\, u(1,1) \,]\, N^*(1,1) + N^*(1,2) \, \alpha \,[\, u(1,1) \,,\, u(1,2) \,]}{K [\, u(1,1) \,]} \right\}$

$= N^*(2,1)\, \beta \,[u(1,1) \,,\, u(2,1) \,]$

[3.27] $R_1 \left\{ 1 - \dfrac{\alpha\,[\,u(1,2)\,,\,u(1,2)\,]\,N^*(1,2) \;+\; N^*(1,1)\,\alpha\,[\,u(1,2)\,,\,u(1,1)\,]}{K\,[\,u(1,2)\,]} \right\}$

$$= \; N^*(2,1)\,\beta\,[u(1,2)\,,\,u(2,1)\,]$$

[3.28] $b \left\{ N^*(1,1)\,\beta\,[\,u(1,1)\,,\,u(2,1)\,] \;+\; N^*(1,2)\,\beta\,[\,u(1,2)\,,\,u(2,1)\,] \right\} = N^*(2,1)\;.$

To illustrate the use of these results, consider the following example. Let

[3.29] $K(u) \;=\; 100\, e^{-\frac{(u-1)^2}{2}}$

[3.30] $\alpha\,[\,u\,,\,u(1,i)\,] \;=\; 1 - \dfrac{[\,u - u(1,i)\,]^2}{16}$

[3.31] $\beta\,[\,u(1,i)\,,\,u] \;=\; 0.15\, e^{-\frac{[\,u(1,i)\,-\,u\,]^2}{2}}$

[3.32] $b \;=\; R_1 \;=\; R_2 \;=\; 0.25\;.$

We begin by seeking an ESS coalition of one prey and one predator. Substituting [3.29]-[3.32] into the necessary conditions [3.11]-[3.14] yields $u(1,1) = u(2,1) = 1$. The equilibrium densities of prey and predators are $N^*(1,1) = 30.77$ and $N^*(2,1) = 1.15$, respectively. At this solution, the prey maximize their carrying capacity and the predator has a strategy which maximizes its capture rate of prey. However, this candidate solution is not an ESS. While the predator strategy maximizes the fitness of the individual predator, the prey's strategy is a local minimum with respect to the fitness of the individual. Any strategy somewhat greater or less than $u(1,1) = 1$ will yield an individual prey higher fitness.

As an alternative, we seek an ESS coalition of two prey and one predator strategy. Substituting [3.29]-[3.32] into necessary conditions [3.23]-[3.28] yields $u(1,1) = 1.8366,$

$u(1,2) = 0.1634$, and $u(2,1) = 1$. The equilibrium densities of prey and predators are $N^*(1,1) = N^*(1,2) = 20.73$ and $N^*(2,1) = 1.10$, respectively. This solution turns out to be an ESS, and we need not seek ESS coalitions with greater numbers of prey or predator strategies.

The ESS can be illustrated graphically using frequency-dependent adaptive landscapes. A frequency-dependent adaptive landscape is a plot of the fitness generating function as a function of u holding the vector of strategies present in the population and vector of population densities constant. There will be a separate landscape for each G function. In a population using the ESS strategies at the equilibrium densities of prey and predator, the prey's ESS strategies are the peaks of the prey's adaptive landscape and the predator's ESS strategy is the peak of the predator's adaptive landscape. This is illustrated in Figure 1 with $(G_1 - 1)$ and $(G_2 - 1)$ plotted versus u. Note that all maximal peaks are at $G_1 = G_2 = 1$.

§4. Discussion

We have modeled predator-prey coevolution as an evolutionary game. The model includes interactions among the prey individuals and the predator individuals and interactions between predator and prey. The strategy of an individual may influence all of these interactions. With respect to trophic structure, the model permits both lateral and vertical coevolutionary responses. Laterally (within a trophic level), the evolution of prey is affected by the competitive interactions between the two prey individuals. Vertically (between trophic levels), the evolutions of predator and prey are influenced by each other's strategies. The specific model used to illustrate the theory results in the evolution of two prey strategies (the predator exerts disruptive selection on the prey). In the absence of the predator, only one prey morph would form the ESS. Interestingly, the two prey strategies, once evolved, can coexist in the absence of the predator. However, this coexistence is not evolutionarily stable. Natural selection will select for the single strategy which, in the absence of the predator, can competitively exclude all other strategies.

The results of the model encourage an expanded view of how predators may contribute to the diversity of a community. Experiments which rely on predator removal and subsequent changes in prey diversity may not be sufficient to detect all circumstances in which predators have exerted disruptive selection. The concept of "predator-mediated coexistence" can be interpreted either ecologically or evolutionarily. Ecologically, the presence of a predator may be necessary to keep one prey species from competitively excluding another (Paine 1966). Evolutionarily, a predator may exert disruptive selection that results in the evolution of additional diversity among the prey (in this paper, the predator alters the prey ESS from a single

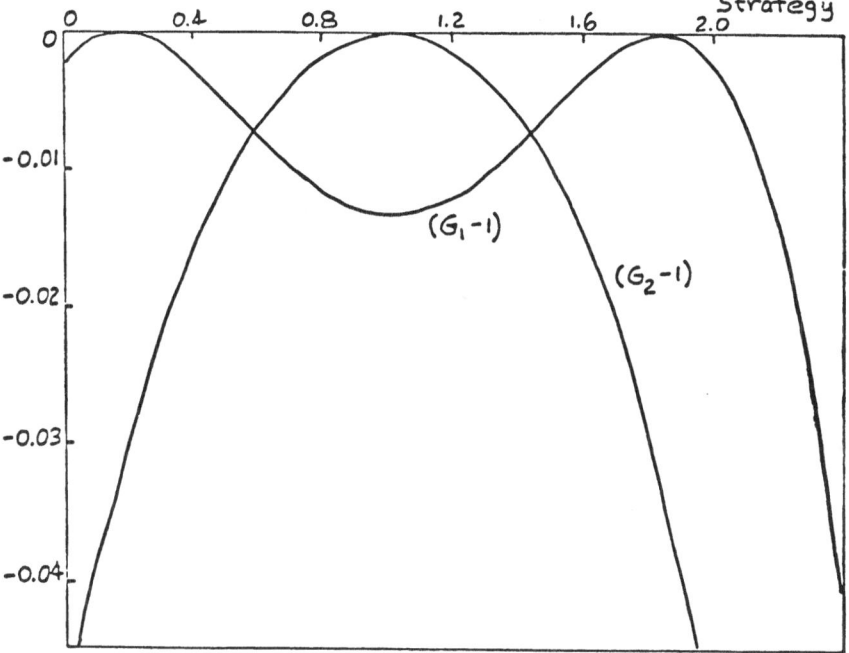

Figure 1. The adaptive landscape for the predator-prey example.

strategy to several strategies). A predator which mediates the coexistence of prey strategies evolutionarily need not do so ecologically or vice versa.

The theory presented in this paper is an extension of our previous theory (Vincent and Brown 1984 , Vincent 1985, Brown and Vincent 1986). Previously, we have assumed that all individuals share the same set of evolutionary possibilities and experience the same ecological consequences of those strategies. In the parlance of the theory, all individuals are represented by the same fitness generating function and draw their strategies from the same strategy set. Evolution within closely related groups such as coyotes, domestic dogs, jackals, and wolves is probably best modeled with a single fitness generating function. However, individuals from distantly related groups, antelopes and canids, do not share the same set of evolutionary strategies and do not have the same ecological consequences of those strategies. Modeling the coevolution between and among such groups requires more than one fitness generating function. We have shown how our previous results may be extended to include any number of fitness generating functions.

The theory permits consideration of ESS's composed of several strategies both within and between fitness generating functions. It is not restricted to modeling predator-prey systems but can be applied to any system that includes organisms with distinct phylogenies and evolutionary potentials. Furthermore, the theory is not restricted to the Lotka-Volterra competition equations or the predator equations used here for illustration. The theory is applicable to any model of population interactions that includes evolutionary variables which can be formulated in terms of fitness generating functions.

Literature Cited

Brown, J.S., and T.L. Vincent. 1987. A theory for the evolutionary game. *Theoretical Population Biology* 31, Number 1.

Brown, J.S., and T.L. Vincent. 1987. Coevolution as an evolutionary game. *Evolution* 41, No. 1.

Paine, R.T. 1966. Food web complexity and species diversity. *The American Naturalist* 100: 65-76.

Rosenzweig, M.L., and W.M. Schaffer. 1978. Homage to the Red Queen. II. Coevolutionary responses to enrichment of exploitation ecosystems. *Theoretical Population Biology* 14: 158-163.

Roughgarden, J. 1983. The theory of coevolution. Pages 33-64 *in* D.J. Futuyma and M. Slatkin (editors). *Coevolution*. Sinauer, Sunderland, Massachusetts.

Schaffer, W.M., and M.L. Rosenzweig. 1978. Homage to the Red. Queen. I. Coevolution of predators and their victims. *Theoretical Population Biology* 14: 135-157.

Vincent, T.L. 1985. Evolutionary games. *Journal of Optimization Theory and Applications* 46: 605-612.

Vincent, T.L., and J.S. Brown. 1984. Stability in an evolutionary game. *Theoretical Population Biology* 26: 408-427.

Vincent, T.L. and Fisher, M.E. 1987. Evolutionarily stable strategies in differential and difference equation models. Manuscript in preparation.

Schaffer, W.M. and M.L. Rothschild, 1978. Iteration in the Pac. Ocean of iteration of systems on their states. Theoretical Population Biology 1-

Vincent, T.L., 1982. Controllability problem of structured frequency. Applications 14: 369-673.

... and ... Brown ... Remarks on ... optimization using feedback. Feedback theory 15: 120-32.

Vincent, T.L. and Mason, M.C., 1987. Controllability and stabilizability of nonlinear. Automatic Mechanical ... in programmer.

Your source for advances in theoretical biology and biomathematics

Journal of

Mathematical Biology

ISSN 0303-6812   Title No. 285

For mathematicians and biologists working in a wide variety of fields – genetics, demography, ecology, neurobiology, epidemiology, morphogenesis, cell biology – the **Journal of Mathematical Biology** publishes:

- papers in which mathematics is used for a better understanding of biological phenomena
- mathematical papers inspired by biological research, and
- papers which yield new experimental data bearing on mathematical models.

The following selection of articles from recent issues reflects the **Journal of Mathematical Biology's** range and scope:

S. J. Merrill: Stochastic models of tumor growth and the probability of elimination by cytotoxic cells. – *H. Aagaard-Hansen, G. F. Veo:* A stochastic discrete generation birth, continuous death population growth model and its approximate solution. – *M. Weiss:* A note on the role of generalized inverse Gaussian distributions of circulatory transit times in pharmacokinetics. – *S. Ellner:* Asymptotic behavior of some stochastic difference equation population models. – *O. Diekmann, H. J. A. M. Heijmans, H. R. Thieme:* On the stability of the cell size distribution. – *A. Hunding:* Bifurcations of nonlinear reaction-diffusion systems in oblate spheroids. – *W. L. Keith, R. H. Rand:* 1:1 and 2:1 phase entrainment in a system of two coupled limit cycle oscillators. – *W. Strittmatter, J. Honerkamp:* Fibrillation of a cardiac region and the tachycardia mode of a two oscillator system. – *V. Comincioli, A. Torelli, C. Poggesi, C. Reggiani:* A four-state cross bridge model for muscle contraction. Mathematical study and validation. – *H. R. Gregorius:* Convergence of genotypic frequencies for differential selfing and positive assortative mating at a biallelic locus. – *J. B. Keller:* Genetic variability due to geographic inhomogeneity.

Subscription information:
To enter your subscription, or to request sample copies, contact Springer-Verlag, Dept. ZSW, Heidelberger Platz 3, D-1000 Berlin 33, W. Germany

Springer-Verlag
Berlin Heidelberg New York
London Paris Tokyo

Springer

Bio-mathematics

Managing Editor: S. A. Levin

Editorial Board: M. Arbib,
H. J. Bremermann, J. Cowan,
W. M. Hirsch, J. Karlin,
J. Keller, K. Krickeberg,
R. C. Lewontin, R. M. May,
J. D. Murray, A. Perelson,
T. Poggio, L. A. Segel

Springer-Verlag
Berlin Heidelberg New York
London Paris Tokyo